AF766981

PRIMAL-DUAL INTERIOR POINT METHODS

PRIMAL-DUAL INTERIOR POINT METHODS

Stephen J. Wright
Argonne National Laboratory
Argonne, Illinois USA

siam
Society for Industrial and Applied Mathematics
Philadelphia

Copyright © 1997 by the Society for Industrial and Applied Mathematics.

10 9 8 7 6 5

All rights reserved. Printed in the United States of America. No part of this book may be reproduced, stored, or transmitted in any manner without the written permission of the publisher. For information, write to the Society for Industrial and Applied Mathematics, 3600 Market Street, 6th Floor, Philadelphia, PA 19104-2688 USA.

Library of Congress Cataloging-in-Publication Data

Wright, Stephen J., 1960-
Primal-dual interior-point methods / Stephen J. Wright.
p. cm.
Includes bibliographical references and index.
1. Linear programming. 2. Mathematical optimization. I. Title.
T57.74.W75 1997
519.7'2 – DC20 96-42071

ISBN: 978-0-898713-82-4

SIAM is a registered trademark.

For Nick, Angela and Charlotte

Contents

Preface

Linear programming has been the dominant paradigm in optimization since Dantzig's development of the simplex method in the 1940s. In 1984, the publication of a paper by Karmarkar started a wave of research into a new class of methods known as interior-point methods, and in the decade since then, *primal-dual algorithms* have emerged as the most important and useful algorithms from this class.

On the theoretical side, the properties of primal-dual methods for linear programming have been quite well understood by researchers since approximately 1994. On the computational side, most interior-point software written since 1990 has been based on a single primal-dual algorithm: Mehrotra's predictor-corrector algorithm. The interesting results are, however, widely scattered in the literature, and most of the relevant papers and reports assume a significant amount of background knowledge on the part of the reader. In this book, we use a simple, unified framework to describe the major results, and we provide a straightforward, self-contained account of the underlying theory.

Primal-dual methods have excellent theoretical properties, good practical performance, and pleasing relationships to earlier fundamental ideas in mathematical programming. They can be extended to wider classes of problems, including convex quadratic programming, monotone linear complementarity, and semidefinite programming, without losing their attributes of simplicity and good practical behavior. Extensions to more difficult classes of problems (even to nonlinear programming) are under way, but the issues become more complicated with the loss of linearity and convexity. These extensions are keeping many researchers busy.

Chapter 1 presents a broad overview of primal-dual methods and should be read first by all nonspecialist readers. In the next two chapters, we fill in some of the background for primal-dual methods with results from the theory of linear programming (Chapter 2) and complexity theory (Chapter 3).

Chapter 2 also contains some general historical background on interior-point methods. Descriptions and full analyses of the most interesting algorithms appear in Chapters 4, 5, and 6. An algorithm based on the primal-dual potential function is the subject of Chapter 4. Chapter 5 describes three algorithms of the path-following genre, while Chapter 6 discusses an infeasible-interior-point method that generalizes the long-step path-following method from the preceding chapter. Issues of fast local convergence and finite termination at an optimal solution are discussed in Chapter 7.

Extensions of primal-dual methods beyond linear programming are outlined in Chapter 8. We discuss monotone linear and nonlinear complementarity problems, convex programming, and semidefinite programming. Techniques for reliable detection of infeasible linear programs, including use of the homogeneous self-dual formulation, are the subject of Chapter 9.

The last two chapters turn to topics of more immediate practical interest. Most current software is based on Mehrotra's predictor-corrector algorithm, which enhances the algorithms of Chapters 4, 5, and 6 with a higher-order search direction and some effective heuristics. These enhancements are motivated and described in Chapter 10. In Chapter 11, we examine the computational issues involved in implementing primal-dual methods. The agenda here is dominated by large sparse systems of linear equations, which must be solved to obtain search directions at each iteration. We also discuss the particular features of some primal-dual codes that have appeared recently in the public domain.

Although this book contains a good deal of analysis, it has a somewhat practical bias. We tend to focus on algorithms that are closely related to practical methods. Primal-dual affine-scaling algorithms are omitted for this reason, despite their fascinating theoretical properties. Within each chapter, the intuitive/descriptive sections of the text are well separated from the more technical sections; the latter can be skipped safely and left to a rainy day.

This book is intended to fulfill a number of needs. It can be used as a text for a course on interior-point methods for linear programming. Selected chapters can be used for a short lecture series within a wider course on optimization or linear programming. The prerequisites are a basic understanding of real analysis and linear algebra and a familiarity with the elements of duality theory for linear programming.

Optimization practitioners and researchers who are not interior-point specialists can use the book as an introduction to some of the major ideas and algorithms. Experts can use it as a reference for the main theoretical results and analytical techniques and also as a guide to aspects of the area with which they are not so familiar.

The field of interior-point methods is an active and competitive one, and the origin of some of the main ideas is sometimes a topic of contention. We have not tried to taxonomize the literature or to trace the development of the various ideas, since these issues are of limited interest to students and nonspecialists. Still, we have tried to point out the key publications, including papers with comprehensive citation lists. No doubt there are some omissions, for which we apologize.

Two sites on the World Wide Web will interest readers of this book. *Interior-Point Methods Online* collects new technical reports in the area, together with other relevant information. Its URL is

`http://www.mcs.anl.gov/home/otc/InteriorPoint/`

The other site is the home page for this book, which contains up-to-date information on software packages, a list of misprints, a feedback column, and so on. The URL is

`http://www.siam.org/books/swright/`

Other Web sites are mentioned in the text as well. The Web is proving to be a valuable adjunct to research activity in this and other fast-moving areas.

I thank Mihai Anitescu, Uri Ascher, Larry Biegler, Joe Czyzyk, Sharon Filipowski, Jean-Pierre Haeberly, Irv Lustig, David Mayne, Michael Overton, Florian Potra, Barry Smith, Yin Zhang, and the anonymous referees of the February 1996 draft for their interest, advice, and extremely valuable comments, which filled numerous gaps in my knowledge. Gail Pieper's assiduous proofreading turned up many infelicities of grammar and logic. As always, her help was invaluable. I am also grateful to Paul Plassmann and Jorge Nocedal for their encouragement during the early stages of this project. Finally, I thank Susan Ciambrano and the staff at SIAM, who made the publication process a sheer delight.

STEPHEN J. WRIGHT
ARGONNE, ILLINOIS

Notation

We summarize the notation and terminology of the book here to avoid confusion and save page ruffling.

$\mathbb{R}^n$	The space of real n-dimensional vectors.
$\mathbb{R}^n_+$	The nonnegative orthant $\{x \in \mathbb{R}^n \mid x \geq 0\}$.
$\mathbb{R}^{m \times n}$	The set of real matrices with dimensions $m \times n$.
$\lVert \cdot \rVert, \lVert \cdot \rVert_2$	Euclidean norm. For $u \in \mathbb{R}^n$, $\lVert u \rVert = \left(\sum_{i=1}^n u_i^2\right)^{1/2}$. For a matrix M, $\lVert M \rVert = \max_{\lVert u \rVert = 1} \lVert Mu \rVert$.
$\lVert \cdot \rVert_1$	The 1-norm. For $u \in \mathbb{R}^n$, $\lVert u \rVert_1 = \sum_{i=1}^n \lvert u_i \rvert$.
$\lVert \cdot \rVert_\infty$	The ∞-norm. For $u \in \mathbb{R}^n$, $\lVert u \rVert_\infty = \max_{i=1,2,\ldots,n} \lvert u_i \rvert$. (Note that $\lVert u \rVert_\infty \leq \lVert u \rVert_2 \leq \lVert u \rVert_1$ for any vector u.)
e	$(1, 1, \ldots, 1)^T$
$\log(\cdot)$	The *natural* logarithm: $\log_e$
x	$\mathbb{R}^n$-vector of primal variables.
λ	$\mathbb{R}^m$-vector of dual variables, that is, Lagrange multipliers for the equality constraints $Ax = b$.
s	$\mathbb{R}^n$-vector of dual slacks, that is, Lagrange multipliers for the bound constraints $x \geq 0$.
A	$\mathbb{R}^{m \times n}$-coefficient matrix for the linear program.
$A_{\cdot i}$	ith column of A.
problem data	The triple (A, b, c), which completely defines the linear programming problem.
L	Number of bits required to store the problem data when all components of (A, b, c) are integers or rational numbers.
$\mathcal{F}$	Primal-dual feasible set:

$\{(x, \lambda, s) \mid Ax = b, A^T\lambda + s = c, (x, s) \geq 0\}$.

$\mathcal{F}^o$ Primal-dual strictly feasible set:
$\{(x, \lambda, s) \mid Ax = b, A^T\lambda + s = c, (x, s) > 0\}$.

duality gap If (x, λ, s) is a feasible point, the duality gap is $c^Tx - b^T\lambda$, which coincides with x^Ts.

μ Duality measure, defined as x^Ts/n.

r_b Primal residual, defined as $Ax - b$.

r_c Dual residual, defined as $A^T\lambda + s - c$.

Z^* Optimal objective at a solution of the linear program. If (x^*, λ^*, s^*) is a primal-dual solution, then $c^Tx^* = Z^* = b^T\lambda^*$. If (x, λ, s) is any feasible point, then $c^Tx \geq Z^* \geq b^T\lambda$.

Ω_P Set of primal solutions x^*.

Ω_D Set of dual solutions (λ^*, s^*).

Ω Set of primal-dual solutions (x^*, λ^*, s^*); $\Omega = \Omega_P \times \Omega_D$.

$\mathcal{B}$ Defined by $i \in \mathcal{B} \subset \{1, 2, \ldots, n\}$ if $x_i^* > 0$ for some primal solution x^*.

$\mathcal{N}$ Defined by $i \in \mathcal{N} \subset \{1, 2, \ldots, n\}$ if $s_i^* > 0$ for some dual slack solution s^*.
$(\mathcal{B} \cup \mathcal{N} = \{1, 2, \ldots, n\}$ and $\mathcal{B} \cap \mathcal{N} = \emptyset$.)

$|\mathcal{B}|$, $|\mathcal{N}|$ The number of elements in $\mathcal{B}$ and $\mathcal{N}$, respectively.

$A_\mathcal{B}$, $A_\mathcal{N}$ Partition of A into columns corresponding to the index sets $\mathcal{B}$ and $\mathcal{N}$.

$x_\mathcal{B}$, $x_\mathcal{N}$ Subvectors of x that correspond to $\mathcal{B}$ and $\mathcal{N}$, respectively. (Similarly for $s_\mathcal{B}$, $s_\mathcal{N}$, $x_\mathcal{B}^*$, etc.)

strictly complementary solution
A solution (x^*, λ^*, s^*) such that $x_\mathcal{B}^* > 0$ and $s_\mathcal{N}^* > 0$.

$\epsilon(A, b, c)$ $\min(\min_{i\in\mathcal{B}} \sup_{x^*\in\Omega_P} x_i^*, \min_{i\in\mathcal{N}} \sup_{(\lambda^*,s^*)\in\Omega_D} s_i^*)$.

$\mathcal{M}$, basis, B A subset of m indices from $\{1, 2, \ldots, n\}$ such that the submatrix $B = [A_{\cdot j}]_{j\in\mathcal{M}}$ is nonsingular and $B^{-1}b \geq 0$.

k (superscript or subscript)
Iteration counter for a sequence, $k = 0, 1, 2, \ldots$.

i, j (subscript)
Usual notation for indices of particular components of a vector or matrix.

(x^k, λ^k, s^k) or (x, λ, s)
Generic iterate of a primal-dual interior-point method. (Iteration counter k often omitted for clarity.)

$(\Delta x, \Delta\lambda, \Delta s)$ Generic primal-dual step, satisfying (1.12) or (1.20), depending on the context.

(x^+, λ^+, s^+) Generic "next point"—the new iterate obtained by taking a step from the current point (x, λ, s).

$F(\cdot)$ A function of (x, λ, s) consisting of the equalities from the Karush–Kuhn–Tucker conditions.

$J(\cdot)$ Jacobian of F.

α, α_k Step length parameter.

unit step A step with $\alpha = 1$.

$(x(\alpha), \lambda(\alpha), s(\alpha))$ $(x, \lambda, s) + \alpha(\Delta x, \Delta\lambda, \Delta s)$.

$(x^k(\alpha), \lambda^k(\alpha), s^k(\alpha))$ $(x^k, \lambda^k, s^k) + \alpha(\Delta x^k, \Delta\lambda^k, \Delta s^k)$.

$\mu(\alpha)$ $x(\alpha)^T s(\alpha)/n$.

$\mu_k(\alpha)$ $x^k(\alpha)^T s^k(\alpha)/n$.

X, X^k $n \times n$ diagonal matrix constructed from the vector x or x^k: $X = \text{diag}(x_1, x_2, \ldots x_n)$.

S, S^k $n \times n$ diagonal matrix constructed from the vector s or s^k: $S = \text{diag}(s_1, s_2, \ldots s_n)$.

D $X^{1/2}S^{-1/2}$.

ΔS, ΔX Diagonal matrices constructed from Δs and Δx, respectively.

v Vector in $\mathbb{R}^n$ with components $(x_i s_i)^{1/2}$.

$v_{\min}$ $\min_i v_i$.

V $\text{diag}(v_1, v_2, \ldots, v_n)$.

r $-v + (n/\rho)\mu V^{-1}e$, for some $\rho > n$.

affine-scaling direction
The pure Newton direction obtained by setting $\sigma = 0$ in (1.12 or (1.20).

centering direction
The direction obtained by setting $\sigma = 1$ in (1.12) or (1.20).

σ	Centering parameter for step computation (1.12), (1.20); restricted to the range $[0,1]$.
Range(B)	Range space of the matrix B.
Null(B^T)	Null space of the matrix B^T. (The fundamental theorem of algebra states that $\text{Null}(B^T) \oplus \text{Range}(B) = \mathbb{R}^n$, where B has n rows.)
$P_{\mathcal{U}}$	Projection into the subspace $\mathcal{U}$.
dist($x, \mathcal{T}$)	Distance from x to the set $\mathcal{T}$, defined as $\inf_{y \in \mathcal{T}} \|x - y\|$.
vertex	If $\mathcal{T}$ is a polyhedron in $\mathbb{R}^n$, a point $x \in \mathcal{T}$ is a vertex if it is the unique minimizer in $\mathcal{T}$ of some linear function.

$\Phi_\rho(x,s) = \rho \log x^T s - \sum_{i=1}^n \log x_i s_i$ for some $\rho \geq n$
The Tanabe–Todd–Ye potential function.

$\mathcal{N}_2(\theta)$	The 2-norm central path neighborhood, defined by $\{(x,\lambda,s) \in \mathcal{F}^o \mid \|XSe - \mu e\| \leq \theta\mu\}$ for given $\theta \in (0,1)$.
$\mathcal{N}_{-\infty}(\gamma)$	The ∞-norm central path neighborhood, defined by $\{(x,\lambda,s) \in \mathcal{F}^o \mid x_i s_i \geq \gamma\mu,\ \forall_i\}$ for given $\gamma \in (0,1)$.
$\mathcal{N}_{-\infty}(\gamma,\beta)$	The infeasible ∞-norm neighborhood, defined by $\{(x,\lambda,s) \mid \|(r_b,r_c)\| \leq [\|(r_b^0,r_c^0)\|/\mu_0]\beta\mu,\ (x,s)>0,\ x_i s_i \geq \gamma\mu,\ \forall_i\}$ for given $\gamma \in (0,1)$, $\beta \geq 1$, and initial point (x^0,λ^0,s^0).
LCP	Monotone linear complementarity problem.
mLCP	Mixed monotone linear complementarity problem.
hLCP	Horizontal monotone linear complementarity problem.

symmetric, skew-symmetric
M is symmetric if $M^T = M$, skew-symmetric if $M^T = -M$.

$\mathcal{S}^n$	The set of $n \times n$ symmetric matrices.
$\mathcal{D}_+$	The set of diagonal matrices with strictly positive diagonal elements.
$\mathbf{u}$	Unit roundoff error (about 10^{-16} for 8-byte floating-point arithmetic).

Chapter 1

Introduction

Linear programming is one of the great success stories of optimization. Since its formulation in the 1930s and 1940s and the development of the simplex algorithm by Dantzig in the mid 1940s, generations of workers in economics, finance, and engineering have been trained to formulate and solve linear programming problems. Even when the situations being modeled are actually nonlinear, linear formulations are favored because the software is highly sophisticated, because the algorithms guarantee convergence to a global minimum, and because uncertainties in the model and data often make it impractical to construct a more elaborate nonlinear model.

The publication in 1984 of Karmarkar's paper [57] was probably the most significant event in linear programming since the discovery of the simplex method. The excitement that surrounded this paper was due partly to a theoretical property of Karmarkar's algorithm—polynomial complexity—and partly to the author's claims of excellent practical performance on large linear programs. These claims were never fully borne out, but the paper sparked a revolution in linear programming research that led to theoretical and computational advances on many fronts. Karmarkar's paper, and earlier work whose importance was recognized belatedly, gave rise to the field of *interior-point methods*, and the years following 1984 saw rapid development and expansion of this new field that continue even today. Theoreticians focused much of their attention on *primal-dual methods*, the elegant and powerful class of interior-point methods that is the subject of this book. Computational experiments, which took place simultaneously with the theoretical development, showed that primal-dual algorithms also performed better than other interior-point methods on practical problems, outperforming even simplex codes on many large problems. A puzzling gap remained between theory and practice, however, since the best practical algorithms differed in a number of respects from the algorithms with nice theoretical properties. Research in the 1990s largely closed this gap, and by 1994, the field had matured.

In this book, we describe the state of the art in primal-dual interior-point methods in their application to linear programming. We start by presenting some background on the duality theory of linear programming and on interior-point methods in Chapter 2. Chapter 3 sketches the theory of algorithmic complexity. The theory for the most successful algorithms in the primal-dual class—potential-reduction, path-following, and infeasible-interior-point algorithms—is presented in Chapters 4, 5, and 6, respectively. In succeeding chapters, we discuss issues of rapid asymptotic convergence (Chapter 7); extensions to more general classes of problems, such as convex quadratic programming (Chapter 8); and detection of infeasible linear programs (Chapter 9). The last two chapters turn to more immediate practical concerns. Chapter 10 discusses Mehrotra's predictor-corrector approach, which, since 1990, has been the basis for most interior-point software. Finally, in Chapter 11, we discuss the issues that arise in implementing primal-dual algorithms. Appendix A contains background information on linear algebra, optimization and complementarity theory, and numerical analysis. Some current software is reviewed in Appendix B.

The remainder of this chapter is devoted to a thumbnail sketch of each of these topics.

Linear Programming

The fundamental properties of a linear programming problem are

a. a vector of real variables, whose optimal values are found by solving the problem;

b. a linear objective function;

c. linear constraints, both inequalities and equalities.

One particular formulation of the linear programming problem—the standard form—is frequently used to describe and analyze algorithms. This form is

$$\min c^T x \text{ subject to } Ax = b,\ x \geq 0, \tag{1.1}$$

where c and x are vectors in $\mathbb{R}^n$, b is a vector in $\mathbb{R}^m$, and A is an $m \times n$ matrix. If x satisfies the constraints $Ax = b$, $x \geq 0$, we call it a *feasible point*; the set of all feasible points is the *feasible set.*

We can convert any linear program into standard form by introducing additional variables—called slack variables and artificial variables—into its formulation.

Associated with any linear program is another linear program called the *dual*, which consists of the same data objects arranged in a different way. The dual for (1.1) is

$$\max\, b^T \lambda \text{ subject to } A^T \lambda + s = c,\ s \geq 0, \tag{1.2}$$

where λ is a vector in $\mathbb{R}^m$ and s is a vector in $\mathbb{R}^n$. We call components of λ the *dual variables*, while s is the vector of *dual slacks.* The dual problem could be stated more compactly by eliminating s from (1.2) and rewriting the constraints as $A^T \lambda \leq c$. However, it turns out to be expedient for the analysis and implementation of interior-point methods to include s explicitly.

The linear programming problem (1.1) often is called the *primal*, to distinguish it from (1.2), and the two problems together are referred to as the *primal-dual pair.*

A *duality theory* that explains the relationship between the two problems (1.1) and (1.2) has been developed. The feasible set and the solution set for the primal tell us a lot about the dual, and vice versa. For instance, given any feasible vectors x for (1.1) and (λ, s) for (1.2), we have that

$$b^T \lambda \leq c^T x. \tag{1.3}$$

In other words, the dual objective gives a lower bound on the primal objective, and the primal gives an upper bound on the dual. The two objective functions coincide at solutions, so that $b^T \lambda^* = c^T x^*$ whenever x^* solves (1.1) and (λ^*, s^*) solves (1.2).

In Chapter 2, we discuss those aspects of the duality theory that have a direct bearing on the design and analysis of interior-point methods. We do not attempt a complete treatment of this fascinating topic, referring the reader instead to the standard reference texts.

Optimality conditions—the algebraic conditions that must be satisfied by solutions of linear programming problems—are derived from first principles and the duality theory in many treatments of linear programming. They also can be stated as special cases of the optimality conditions for general constrained optimization, known as the Karush–Kuhn–Tucker (or KKT) conditions. We state the KKT conditions in Appendix A, in a form that is sufficiently general for the purposes of this book (see Theorem A.1). The optimality conditions for the primal problem (1.1) are obtained by specializing Theorems A.1 and A.2:

The vector $x^* \in \mathbb{R}^n$ *is a solution of* (1.1) *if and only if there exist vectors* $s^* \in \mathbb{R}^n$ *and* $\lambda^* \in \mathbb{R}^m$ *for which the following conditions hold for* $(x, \lambda, s) = (x^*, \lambda^*, s^*)$*:*

$$\begin{aligned} A^T\lambda + s &= c, && (1.4a)\\ Ax &= b, && (1.4b)\\ x_i s_i &= 0, \quad i = 1, 2, \ldots, n, && (1.4c)\\ (x, s) &\geq 0. && (1.4d) \end{aligned}$$

In the terminology of general constrained optimization, the vectors λ and s are *Lagrange multipliers* for the constraints $Ax = b$ and $x \geq 0$, respectively. Condition (1.4c) implies that for each index $i = 1, 2, \ldots, n$, one of the components x_i or s_i must be zero. This condition is known as the *complementarity condition*, since it implies that the nonzeros of x and s appear in complementary locations.

Note that in (1.4) the Lagrange multipliers are denoted by the same symbols—λ and s—that we use for the unknowns in the dual problem (1.2). This choice is not an accident, for if we apply Theorems A.1 and A.2 to the dual problem, we find that the optimality conditions make use of exactly the same conditions (1.4). This is the crucial observation that defines the relationship between the primal and dual problems. Formally, we state the dual optimality conditions as follows:

The vector $(\lambda^*, s^*) \in \mathbb{R}^m \times \mathbb{R}^n$ *is a solution of* (1.2) *if and only if there exists a vector* $x^* \in \mathbb{R}^n$ *such that the conditions* (1.4) *hold for* $(x, \lambda, s) = (x^*, \lambda^*, s^*)$.

By examining the conditions (1.4) from both the primal and the dual viewpoints, we conclude that a vector (x^*, λ^*, s^*) solves the system (1.4) if and only if x^* solves the primal problem (1.1) and (λ^*, s^*) solves the dual problem (1.2). The vector (x^*, λ^*, s^*) is called a *primal-dual solution.*

Primal-Dual Methods

This book is about *primal-dual interior-point methods.* These methods find primal-dual solutions (x^*, λ^*, s^*) by applying variants of Newton's method to the three equality conditions in (1.4) and modifying the search directions and step lengths so that the inequalities $(x, s) \geq 0$ are satisfied *strictly* at every iteration. It is precisely the innocuous-looking bounds on x and s that give rise to all the complications in the design, analysis, and implementation of the methods described in this book.

Let us restate the optimality conditions (1.4) in a slightly different form by means of a mapping F from $\mathbb{R}^{2n+m}$ to $\mathbb{R}^{2n+m}$:

$$F(x, \lambda, s) = \begin{bmatrix} A^T\lambda + s - c \\ Ax - b \\ XSe \end{bmatrix} = 0, \tag{1.5a}$$

$$(x, s) \geq 0, \tag{1.5b}$$

where

$$X = \text{diag}(x_1, x_2, \ldots, x_n), \qquad S = \text{diag}(s_1, s_2, \ldots, s_n), \tag{1.6}$$

and $e = (1, 1, \ldots, 1)^T$. Note that F is actually linear in its first two terms $Ax - b$, $A^T\lambda + s - c$, and only mildly nonlinear in the remaining term XSe.

All primal-dual methods generate iterates (x^k, λ^k, s^k) that satisfy the bounds (1.5b) strictly, that is, $x^k > 0$ and $s^k > 0$. This property is the origin of the term *interior-point*. By respecting these bounds, the methods avoid spurious solutions, which are points that satisfy $F(x, \lambda, s) = 0$ but not $(x, s) \geq 0$. Spurious solutions abound, and none of them gives any useful information about solutions of (1.1) or (1.2), so it is best to exclude them altogether from the region of search. Most interior-point methods actually require the iterates to be *strictly feasible*; that is, each (x^k, λ^k, s^k) must satisfy the linear equality constraints for the primal and dual problems. If we define the primal-dual *feasible set* $\mathcal{F}$ and *strictly feasible set* $\mathcal{F}^o$ by

$$\mathcal{F} = \{(x, \lambda, s) \mid Ax = b, A^T\lambda + s = c, (x, s) \geq 0\}, \tag{1.7a}$$

$$\mathcal{F}^o = \{(x, \lambda, s) \mid Ax = b, A^T\lambda + s = c, (x, s) > 0\}, \tag{1.7b}$$

the strict feasibility condition can be written concisely as

$$(x^k, \lambda^k, s^k) \in \mathcal{F}^o.$$

Like most iterative algorithms in optimization, primal-dual interior-point methods have two basic ingredients: a procedure for determining the step and a measure of the desirability of each point in the search space. As mentioned earlier, the search direction procedure has its origins in Newton's method for the nonlinear equations (1.5a). Newton's method forms a linear model for F around the current point and obtains the search direction $(\Delta x, \Delta\lambda, \Delta s)$ by solving the following system of linear equations:

$$J(x,\lambda,s)\begin{bmatrix}\Delta x\\ \Delta\lambda\\ \Delta s\end{bmatrix} = -F(x,\lambda,s),$$

where J is the Jacobian of F. If the current point is strictly feasible (that is, $(x,\lambda,s) \in \mathcal{F}^o$), the Newton step equations become

$$\begin{bmatrix}0 & A^T & I\\ A & 0 & 0\\ S & 0 & X\end{bmatrix}\begin{bmatrix}\Delta x\\ \Delta\lambda\\ \Delta s\end{bmatrix} = \begin{bmatrix}0\\ 0\\ -XSe\end{bmatrix}. \tag{1.8}$$

A full step along this direction usually is not permissible, since it would violate the bound $(x,s) \geq 0$. To avoid this difficulty, we perform a line search along the Newton direction so that the new iterate is

$$(x,\lambda,s) + \alpha(\Delta x, \Delta\lambda, \Delta s)$$

for some line search parameter $\alpha \in (0,1]$. Unfortunately, we often can take only a small step along the direction ($\alpha \ll 1$) before violating the condition $(x,s) > 0$; hence, the pure Newton direction (1.8) often does not allow us to make much progress toward a solution.

Primal-dual methods modify the basic Newton procedure in two important ways:

1. They bias the search direction toward the interior of the nonnegative orthant $(x,s) \geq 0$ so that we can move further along the direction before one of the components of (x,s) becomes negative.

2. They keep the components of (x,s) from moving "too close" to the boundary of the nonnegative orthant. Search directions computed from points that are close to the boundary tend to be distorted, and little progress can be made along them.

We consider these modifications in turn.

The Central Path

The central path $\mathcal{C}$ is an arc of strictly feasible points that plays a vital role in the theory of primal-dual algorithms. It is parametrized by a scalar $\tau > 0$, and each point $(x_\tau, \lambda_\tau, s_\tau) \in \mathcal{C}$ solves the following system:

$$
\begin{aligned}
A^T\lambda + s &= c, && && (1.9a)\\
Ax &= b, && && (1.9b)\\
x_i s_i &= \tau, && i = 1, 2, \dots, n, && (1.9c)\\
(x, s) &> 0. && && (1.9d)
\end{aligned}
$$

These conditions differ from the KKT conditions only in the term τ on the right-hand side of (1.9c). Instead of the complementarity condition (1.4c), we require that the pairwise products $x_i s_i$ have the same value for all indices i. From (1.9), we can define the central path as

$$\mathcal{C} = \{(x_\tau, \lambda_\tau, s_\tau) \,|\, \tau > 0\}.$$

Another way of defining $\mathcal{C}$ is to use the notation introduced in (1.5) and write

$$F(x_\tau, \lambda_\tau, s_\tau) = \begin{bmatrix} 0 \\ 0 \\ \tau e \end{bmatrix}, \qquad (x_\tau, s_\tau) > 0. \tag{1.10}$$

We show in Chapter 2 that $(x_\tau, \lambda_\tau, s_\tau)$ is defined uniquely for each $\tau > 0$ if and only if $\mathcal{F}^o$ is nonempty. Hence, the entire path $\mathcal{C}$ is well defined.

The equations (1.9) approximate (1.4) more and more closely as τ goes to zero. If $\mathcal{C}$ converges to anything as $\tau \downarrow 0$, it must converge to a primal-dual solution of the linear program. The central path thus guides us to a solution along a route that steers clear of spurious solutions by keeping all the pairwise products $x_i s_i$ strictly positive and decreasing them to zero at the same rate.

Most primal-dual algorithms take Newton steps toward points on $\mathcal{C}$ for which $\tau > 0$, rather than pure Newton steps for F. Since these steps are biased toward the interior of the nonnegative orthant defined by $(x, s) \geq 0$, it usually is possible to take longer steps along them than along the pure Newton steps for F before violating the positivity condition. To describe the biased search direction, we introduce a *centering parameter* $\sigma \in [0, 1]$ and a *duality measure* μ defined by

$$\mu = \frac{1}{n}\sum_{i=1}^{n} x_i s_i = x^T s/n, \tag{1.11}$$

which measures the average value of the pairwise products $x_i s_i$. The generic step equations are then

$$\begin{bmatrix} 0 & A^T & I \\ A & 0 & 0 \\ S & 0 & X \end{bmatrix} \begin{bmatrix} \Delta x \\ \Delta \lambda \\ \Delta s \end{bmatrix} = \begin{bmatrix} 0 \\ 0 \\ -XSe + \sigma\mu e \end{bmatrix}. \tag{1.12}$$

The step $(\Delta x, \Delta\lambda, \Delta s)$ is a Newton step toward the point $(x_{\sigma\mu}, \lambda_{\sigma\mu}, s_{\sigma\mu}) \in \mathcal{C}$, at which the pairwise products $x_i s_i$ are all equal to $\sigma\mu$. In contrast, the step (1.8) aims directly for the point at which the KKT conditions (1.4) are satisfied.

If $\sigma = 1$, the equations (1.12) define a *centering direction*, a Newton step toward the point $(x_\mu, \lambda_\mu, s_\mu) \in \mathcal{C}$, at which all the pairwise products $x_i s_i$ are identical to μ. Centering directions are usually biased strongly toward the interior of the nonnegative orthant and make little, if any, progress in reducing μ. However, by moving closer to $\mathcal{C}$, they set the scene for substantial progress on the next iteration. (Since the next iteration starts near $\mathcal{C}$, it will be able to take a relatively long step without leaving the nonnegative orthant.) At the other extreme, the value $\sigma = 0$ gives the standard Newton step (1.8), sometimes known as the *affine-scaling direction* for reasons to be discussed later. Many algorithms use intermediate values of σ from the open interval $(0, 1)$ to trade off between the twin goals of reducing μ and improving centrality.

A Primal-Dual Framework

With these basic concepts in hand, we can define Framework PD, a general framework for primal-dual algorithms. Most methods in this book are special cases of Framework PD.

Framework PD
Given $(x^0, \lambda^0, s^0) \in \mathcal{F}^o$
for $k = 0, 1, 2, \ldots$
solve

$$\begin{bmatrix} 0 & A^T & I \\ A & 0 & 0 \\ S^k & 0 & X^k \end{bmatrix} \begin{bmatrix} \Delta x^k \\ \Delta \lambda^k \\ \Delta s^k \end{bmatrix} = \begin{bmatrix} 0 \\ 0 \\ -X^k S^k e + \sigma_k \mu_k e \end{bmatrix}, \tag{1.13}$$

where $\sigma_k \in [0, 1]$ and $\mu_k = (x^k)^T s^k / n$;

set

$$(x^{k+1}, \lambda^{k+1}, s^{k+1}) \leftarrow (x^k, \lambda^k, s^k) + \alpha_k(\Delta x^k, \Delta\lambda^k, \Delta s^k), \tag{1.14}$$

choosing α_k so that $(x^{k+1}, s^{k+1}) > 0$.

end (for).

Path-Following Methods

A path-following algorithm explicitly restricts the iterates to a neighborhood of the central path $\mathcal{C}$ and follows $\mathcal{C}$ to a solution of the linear program. The neighborhood excludes points (x, s) that are too close to the boundary of the nonnegative orthant. Therefore, search directions calculated from any point in the neighborhood make at least minimal progress toward the solution set.

Recall that a key ingredient of any optimization algorithm is a measure of the desirability of each point in the search space. In path-following algorithms, the duality measure μ defined by (1.11) fills this role. The duality measure μ_k is forced to zero as $k \to \infty$, so the iterates (x^k, λ^k, s^k) come closer and closer to satisfying the KKT conditions (1.4).

The two most interesting neighborhoods of $\mathcal{C}$ are the so-called 2-norm neighborhood $\mathcal{N}_2(\theta)$ defined by

$$\mathcal{N}_2(\theta) = \{(x, \lambda, s) \in \mathcal{F}^o \mid \|XSe - \mu e\|_2 \leq \theta\mu\} \tag{1.15}$$

for some $\theta \in (0, 1)$, and the one-sided ∞-norm neighborhood $\mathcal{N}_{-\infty}(\gamma)$ defined by

$$\mathcal{N}_{-\infty}(\gamma) = \{(x, \lambda, s) \in \mathcal{F}^o \mid x_i s_i \geq \gamma\mu \quad \text{all } i = 1, 2, \ldots, n\} \tag{1.16}$$

for some $\gamma \in (0, 1)$. (Typical values of the parameters are $\theta = 0.5$ and $\gamma = 10^{-3}$.) If a point lies in $\mathcal{N}_{-\infty}(\gamma)$, each pairwise product $x_i s_i$ must be at least some small multiple γ of their average value μ. This requirement is actually quite modest, and we can make $\mathcal{N}_{-\infty}(\gamma)$ encompass most of the feasible region $\mathcal{F}$ by choosing γ close to zero. The $\mathcal{N}_2(\theta)$ neighborhood is more restrictive, since certain points in $\mathcal{F}^o$ do not belong to $\mathcal{N}_2(\theta)$ no matter how close θ is chosen to its upper bound of 1.

By keeping all iterates inside one or another of these neighborhoods, path-following methods reduce all the pairwise products $x_i s_i$ to zero at more or less the same rate.

Path-following methods are akin to homotopy methods for general nonlinear equations, which also define a path to be followed to the solution.

Traditional homotopy methods stay in a tight tubular neighborhood of their path, making incremental changes to the parameter and chasing the homotopy path all the way to a solution. For primal-dual methods, this neighborhood is conical rather than tubular, and it tends to be broad and loose for larger values of the duality measure μ. It narrows as $\mu \to 0$, however, because of the positivity requirement $(x, s) > 0$.

Some path-following methods choose conservative values for the centering parameter σ (that is, σ only slightly less than 1) so that unit steps can be taken along the resulting direction from (1.12) without leaving the chosen neighborhood. These methods, which are known as *short-step* path-following methods, make only slow progress toward the solution because a restrictive $\mathcal{N}_2$ neighborhood is needed to make them work.

Better results are obtained with the *predictor-corrector* method, which uses two $\mathcal{N}_2$ neighborhoods, nested one inside the other. Every second step of this method is a "predictor" step, which starts in the inner neighborhood and moves along the affine-scaling direction (computed by setting $\sigma = 0$ in (1.12)) to the boundary of the outer neighborhood. The gap between neighborhood boundaries is wide enough to allow this step to make significant progress in reducing μ. Between these predictor steps, the algorithm takes "corrector" steps (computed with $\sigma = 1$ and $\alpha = 1$), which take it back inside the inner neighborhood to prepare for the next predictor step. The predictor-corrector algorithm converges superlinearly, as we show in Chapter 7.

Long-step path-following methods make less conservative choices of σ than their short-step cousins. As a consequence, they need to perform a line search along $(\Delta x, \Delta\lambda, \Delta s)$ to avoid leaving the chosen neighborhood. By making judicious choices of σ, however, long-step methods can make much more rapid progress than short-step methods, particularly when the $\mathcal{N}_{-\infty}$ neighborhood is used.

All three methods are discussed and analyzed in Chapter 5.

Potential-Reduction Methods

Potential-reduction methods take steps of the same form as do path-following methods, but they do not explicitly follow $\mathcal{C}$ and can be motivated independently of it. They use a logarithmic potential function to measure the worth of each point in $\mathcal{F}^o$ and aim to achieve a certain fixed reduction in this function at each iteration. The primal-dual potential function, which we denote generically by Φ, usually has two important properties:

$$\Phi \rightarrow \infty \text{ if } x_i s_i \to 0 \text{ for some } i, \text{ but } \mu = x^T s/n \not\to 0, \qquad (1.17a)$$

$$\Phi \rightarrow -\infty \text{ if and only if } (x, \lambda, s) \rightarrow \Omega, \tag{1.17b}$$

where Ω is the set of primal-dual solutions to (1.1). The first property (1.17a) stops any one of the pairwise products $x_i s_i$ from approaching zero independently of the others and therefore keeps the iterates away from the boundary of the nonnegative orthant. The second property (1.17b) relates Φ to the solution set Ω. If our algorithm forces Φ to $-\infty$, then (1.17b) ensures that the sequence approaches the solution set.

The most interesting primal-dual potential function is the Tanabe–Todd–Ye function Φ_ρ, defined by

$$\Phi_\rho(x, s) = \rho \log x^T s - \sum_{i=1}^{n} \log x_i s_i \tag{1.18}$$

for some parameter $\rho > n$. Like all algorithms based on Framework PD, potential-reduction algorithms obtain their search directions by solving (1.13) for some $\sigma_k \in (0, 1)$. The step length α_k is chosen to approximately minimize Φ_ρ along the computed direction. In Chapter 4, we see that cookbook choices of σ_k and α_k are sufficient to guarantee constant reduction in Φ_ρ at every iteration. Hence, Φ_ρ will approach $-\infty$, forcing convergence. Smarter choices of σ_k and α_k (adaptive and heuristic) are also covered by the theory, provided that they yield at least as much reduction in Φ_ρ as do the cookbook values.

Infeasible Starting Points

So far, we have assumed that the starting point (x^0, λ^0, s^0) is strictly feasible and, in particular, that it satisfies the linear equations $Ax^0 = b$, $A^T\lambda^0 + s^0 = c$. All subsequent iterates also respect these constraints, because of the zero right-hand side terms in (1.13).

For most problems, however, a strictly feasible starting point is difficult to find. This task can always be made trivial by reformulating the problem, but the reformulation sometimes introduces distortions that can make the problem harder to solve. An alternative approach is provided by *infeasible-interior-point* methods, which require nothing more of the starting point than positivity of x^0 and s^0. The search direction needs to be modified so that it moves closer to feasibility as well as to centrality, but only a slight change is needed to the step equation (1.12). If we define the residuals for the two linear equations as

$$r_b = Ax - b, \qquad r_c = A^T\lambda + s - c, \tag{1.19}$$

the step equation becomes

$$\begin{bmatrix} 0 & A^T & I \\ A & 0 & 0 \\ S & 0 & X \end{bmatrix} \begin{bmatrix} \Delta x \\ \Delta \lambda \\ \Delta s \end{bmatrix} = \begin{bmatrix} -r_c \\ -r_b \\ -XSe + \sigma\mu e \end{bmatrix}. \tag{1.20}$$

The search direction is still a Newton step toward the point $(x_{\sigma\mu}, \lambda_{\sigma\mu}, s_{\sigma\mu}) \in \mathcal{C}$. It tries to bite off all the infeasibility in the equality constraints in a single step. If a full step is ever taken (that is, $\alpha = 1$), the residuals r_b and r_c become zero, and all subsequent iterates remain strictly feasible.

In Chapter 6, we give a complete description and analysis of a path-following infeasible-interior-point method, which is a variant of the long-step feasible method from Chapter 5. It confines all iterates to a central path neighborhood like $\mathcal{N}_{-\infty}(\gamma)$, extended to allow violation of the linear equality conditions. In this extended neighborhood, the residual norms $\|r_b\|$ and $\|r_c\|$ are bounded by a constant multiple of the duality measure μ. By squeezing μ to zero, we also force r_b and r_c to zero so that we approach complementarity and feasibility simultaneously.

Superlinear Convergence

We now return to the place from which we started, where we motivated primal-dual algorithms as modifications of Newton's method applied to $F(x, \lambda, s) = 0$. Because of the modifications, the search direction often contains a centering term (when $\sigma > 0$), and the step length α is often smaller than its "natural" value of 1. The modifications lead to nice global convergence properties and good practical performance, but they interfere with the best-known characteristic of Newton's method: fast asymptotic convergence. Fortunately, it is possible to design algorithms that recover this important property without sacrificing the benefits of the modified algorithm. When the iterates get close enough to the solution set, there is less danger of the pure Newton step (1.8) being attracted to spurious solutions. By relaxing some of the restrictions and modifications that are applied on earlier iterations, we can allow the algorithm to take steps more like the pure Newton step and therefore achieve a fast asymptotic rate of convergence. We do not abandon the restrictions altogether, however, since we wish to retain the excellent global convergence properties.

The superlinear convergence theory for interior-point methods has a slightly different character from the theory for nonlinear equations. For a general nonlinear system $F(z) = 0$, where the mapping $F : \mathbb{R}^N \rightarrow \mathbb{R}^N$ is continuously differentiable, the sequence $\{z^k\}$ generated by Newton's method converges superlinearly to a solution z^* if the solution is *nondegenerate*; that is, the Jacobian $J(z^*)$ is nonsingular. In the interior-point setting, we can achieve superlinear convergence even when the Jacobian of F approaches a singular limit and even when the solution is not unique! It turns out, as we show in Chapter 7, that the constraints $(x, s) \geq 0$ and our insistence on staying away from the boundary of the nonnegative orthant provide the extra structure that we need for superlinearity.

Extensions

Primal-dual methods for linear programming can be extended to wider classes of problems. Extensions to the monotone linear complementarity problem (LCP) and convex quadratic programming (QP) are straightforward. The monotone LCP—the qualifier "monotone" is implicit throughout this book—is the problem of finding vectors x and s in $\mathbb{R}^n$ that satisfy the following conditions:

$$s = Mx + q, \qquad (x, s) \geq 0, \qquad x^T s = 0, \tag{1.21}$$

where M is a positive semidefinite $n \times n$ matrix and $q \in \mathbb{R}^n$. The similarity between (1.21) and the KKT conditions (1.4) is obvious: the last two conditions in (1.21) correspond to (1.4d) and (1.4c), respectively, while the condition $s = Mx + q$ is similar to the equations (1.4a) and (1.4b).

In many situations, it is more convenient to use a formulation of the LCP that differs slightly from (1.21). In Chapter 8, we discuss two formulations that are more flexible than (1.21) but are equivalent to this basic formulation in a strong sense. The LCP deserves the attention we pay to it in Chapter 8 because it is such a versatile tool for formulating and solving a wide range of problems via primal-dual methods.

In convex QP, we minimize a convex quadratic objective subject to linear constraints. A convex quadratic generalization of the standard form linear program (1.1) is

$$\min c^T x + \tfrac{1}{2} x^T Q x \text{ subject to } Ax = b,\ x \geq 0, \tag{1.22}$$

where Q is a symmetric $n \times n$ positive semidefinite matrix. The KKT conditions for this problem are similar to (1.4) and (1.21). In fact, we can show that any LCP can be formulated as a convex quadratic program, and vice versa.

We also discuss extensions to a new class known as *semidefinite programming*. The name comes from the fact that some of the variables in these problems are square matrices that are constrained to be positive semidefinite. This class, which has been the topic of concentrated research since 1993, has applications in many areas, including control theory and combinatorial optimization.

The monotone nonlinear complementarity problem is obtained by replacing the first (linear) condition in (1.21) by a nonlinear monotone function. That is, we look for vectors x and s in $\mathbb{R}^n$ such that

$$s = f(x), \qquad (x, s) \geq 0, \qquad x^T s = 0,$$

where f is a smooth function with the property that $(x_1 - x_0)^T (f(x_1) - f(x_0)) \geq 0$ for all x_0 and x_1 in some domain of interest. Conceptually, the approach is not much different from the one in Framework PD. The central path is defined analogously to (1.9), and the search directions are still Newton steps toward points on this path. Handling of infeasibility becomes a more prominent issue, however, since it is no longer obvious how we can retain feasibility of the iterates (x^k, s^k) with respect to the nonlinear constraint $s^k = f(x^k)$.

In Chapter 8, we show how primal-dual methods can be extended to these more general problems. In the case of convex QP and LCP, the algorithms and their analysis are fairly straightforward generalizations of the linear programming techniques of Chapters 4, 5, and 6. On the other hand, extensions to nonconvex problems (especially nonlinear programming problems) raise many new and challenging issues that are beyond the scope of this book.

Mehrotra's Predictor-Corrector Algorithm

Most existing interior-point codes for general-purpose linear programming problems are based on Mehrotra's predictor-corrector algorithm [88]. The two key features of this algorithm are

a. addition of a *corrector step* to the search direction of Framework PD so that the algorithm more closely follows a trajectory to the solution set Ω;

b. adaptive choice of the centering parameter σ.

Mehrotra's method can be motivated by considering path-following algorithms for trajectories in $\mathbb{R}^n \times \mathbb{R}^m \times \mathbb{R}^n$. The central path $\mathcal{C}$ (1.9) is one such trajectory. As we show in Chapter 2, $\mathcal{C}$ is a well-defined curve that terminates at the solution set Ω. A modified trajectory that leads from the current

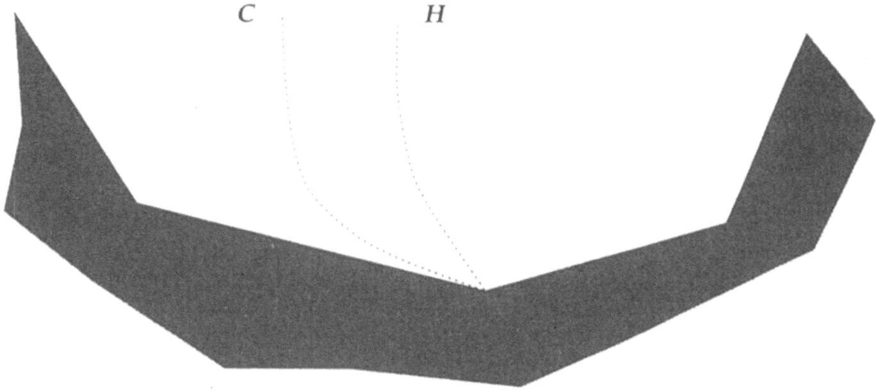

Figure 1.1. Central path $\mathcal{C}$ and a trajectory $\mathcal{H}$ from the current (noncentral) point (x, λ, s) to the solution set Ω.

point (x, λ, s) to Ω is of more immediate interest in designing algorithms, however, since the current iterate rarely lies on the central path and usually is not even feasible. These modified trajectories can be defined in a number of different ways, as we show in Chapter 10. Their common features are that they consist of points $(\hat{x}_\tau, \hat{\lambda}_\tau, \hat{s}_\tau)$ for $\tau \in [0, 1)$, with $(\hat{x}_0, \hat{\lambda}_0, \hat{s}_0) = (x, \lambda, s)$, and if the limit $\lim_{\tau\uparrow 1} (\hat{x}_\tau, \hat{\lambda}_\tau, \hat{s}_\tau)$ exists, it belongs to the solution set Ω. A trajectory $\mathcal{H}$ with these properties is depicted in Figure 1.1.

Algorithms from Framework PD can be thought of as first-order methods in that they find the tangent to a trajectory like $\mathcal{H}$ and perform a line search along it. Mehrotra's algorithm takes the next logical step of calculating the *curvature* of $\mathcal{H}$ at the current point as well as the tangent, thereby obtaining a second-order approximation to the trajectory. The curvature, which is equivalent to the corrector step mentioned above, can be obtained at relatively low cost. Computational experience shows that the extra cost is easily justified because it usually leads to a significant reduction in iteration count over methods based strictly on Framework PD.

The second important feature of Mehrotra's algorithm is that it chooses the centering parameter σ_k *adaptively*, in contrast to algorithms from Framework PD, which assign a value to σ_k prior to calculating the search direction. At each iteration, Mehrotra's algorithm first calculates the affine-scaling direction and assesses its usefulness as a search direction. If this direction yields a large reduction in μ without violating the positivity condition $(x, s) > 0$, the algorithm concludes that little centering is needed, so it chooses σ_k close to zero and calculates a centered search direction with this small value. If the affine-scaling direction is not so useful, the algorithm enforces a larger amount of centering by choosing a value of σ_k closer to 1. Computation of the centered direction and the corrector step can be combined, so adaptive centering does not add further to the cost of each iteration.

Mehrotra's method also incorporates a number of other algorithmic devices that contribute to its practical success, including the use of different step lengths for the primal and dual variables and a heuristic choice of the step lengths that is designed to speed the asymptotic convergence. We discuss Mehrotra's algorithm in Chapter 10.

Linear Algebra Issues

Most of the computational effort in implementations of primal-dual methods is taken up in solving linear systems of the forms (1.12) and (1.20). The coefficient matrix in these systems is usually large and sparse, since the constraint matrix A is itself large and sparse in most applications. The special structure in the step equations (1.12) and (1.20) allows us to reformulate them as systems with more compact symmetric coefficient matrices, which are easier and cheaper to factor than the original form.

The reformulation procedures are simple, as we show by applying them to the system (1.20). Since the current point (x, λ, s) has x and s strictly positive, the diagonal matrices X and S are nonsingular. Hence, by eliminating Δs from (1.20), we obtain the following equivalent system:

$$\begin{aligned} \begin{bmatrix} 0 & A \\ A^T & -D^{-2} \end{bmatrix} \begin{bmatrix} \Delta\lambda \\ \Delta x \end{bmatrix} &= \begin{bmatrix} -r_b \\ -r_c + s - \sigma\mu X^{-1}e \end{bmatrix}, & \text{(1.23a)} \\ \Delta s &= -s + \sigma\mu X^{-1}e - X^{-1}S\Delta x, & \text{(1.23b)} \end{aligned}$$

where we have introduced the notation

$$D = S^{-1/2}X^{1/2}. \tag{1.24}$$

This form of the step equations usually is known as the *augmented system.* Since the matrix $X^{-1}S$ is also diagonal and nonsingular, we can go a step further, eliminating Δx from (1.23a) to obtain another equivalent form:

$$\begin{aligned} AD^2A^T\Delta\lambda &= -r_b + A(-S^{-1}Xr_c + x - \sigma\mu S^{-1}e), & \text{(1.25a)} \\ \Delta s &= -r_c - A^T\Delta\lambda, & \text{(1.25b)} \\ \Delta x &= -x + \sigma\mu S^{-1}e - S^{-1}X\Delta s. & \text{(1.25c)} \end{aligned}$$

This form often is called the *normal equations* form because, in the case of $r_b = 0$, the system (1.25a) reduces to the normal equations for a linear least squares problem with coefficient matrix DA^T.

Most implementations are based on the normal equations form; they apply a direct sparse Cholesky algorithm to factor the matrix AD^2A^T. General-purpose sparse Cholesky software can be used to perform this operation; for example, two of the codes that we discuss in Appendix B make use of the sparse Cholesky code of Ng and Peyton [103]. A few modifications are needed to standard Cholesky codes, however, because AD^2A^T may be ill conditioned or singular. Ill conditioning is often observed during the final stages of a primal-dual algorithm, when the elements of the diagonal weighting matrix D^2 take on both huge and tiny values.

It becomes inefficient to formulate and solve the system (1.25a) when $AS^{-1}XA^T$ is much denser than A. This situation arises when A has one or more dense columns, and it is not uncommon in practice. We can modify the normal equations strategy by excluding the dense columns from the matrix-matrix product $AS^{-1}XA^T$ and correcting for this omission in a subsequent calculation. Details are given in Chapter 11.

The formulation (1.23) has received less attention than (1.25), mainly because algorithms and software for factoring sparse symmetric indefinite matrices are more complicated, slower, and less readily available than sparse Cholesky algorithms. This situation is changing, however, as the result of an increasing recognition that symmetric indefinite matrices are important in many contexts besides (1.23a) (see, for instance, Vavasis [140]). The formulation (1.23) is cleaner and more flexible than (1.25) in a number of respects: The difficulty caused by dense columns in A does not arise, and free variables (that is, components of x with no explicit lower or upper bounds) can be handled without resorting to the various artificial devices needed in the normal equations form.

More details about solving both formulations (1.23) and (1.25) and about other issues associated with primal-dual implementations can be found in Chapter 11.

Karmarkar's Algorithm

For many nonspecialists, the name Karmarkar is more or less synonymous with the whole field of interior-point methods. Hence, a few words about Karmarkar's method [57] and its place in history are in order.

Although the practical efficiency of the simplex method was appreciated from the time it was discovered, theoreticians remained uneasy because the method is not guaranteed to perform well on every linear program. A famous example due to Klee and Minty [61] requires 2^n simplex iterations to solve a linear program in $\mathbb{R}^n$. By contrast, other computational problems such as sorting and searching were known to be solvable by polynomial algorithms—algorithms whose run time is, at worst, a polynomial function of the amount of storage needed for the problem data.

The first polynomial algorithm for linear programming, Khachiyan's ellipsoid algorithm [60], was a computational disappointment. Its convergence is much too slow, it is not robust in the presence of rounding errors, and it requires a large, dense $n \times n$ matrix to be stored and updated at every iteration. It is not competitive with the simplex method on any but the smallest problems.

Karmarkar's algorithm was announced in 1984, five years after the ellipsoid method. It too is polynomial, requiring $O(n \log 1/\epsilon)$ iterations to identify a primal feasible point x such that $c^T x$ is within ϵ of its optimal value for (1.1). Karmarkar's is a *primal* algorithm; that is, it is described, motivated, and implemented purely in terms of the primal problem (1.1) without reference to the dual. At each iteration, Karmarkar's algorithm performs a projective transformation on the primal feasible set that maps the current iterate x^k to the center of the set and takes a step in the feasible steepest descent direction for the transformed space. Progress toward optimality is measured by a logarithmic potential function, as discussed in Chapter 2. Nice descriptions of the algorithm can be found in Karmarkar's original paper [57] and in Fletcher [30].

Karmarkar's method falls outside the scope of this book, and, in any case, its practical performance does not appear to match the most efficient primal-dual methods. The algorithms we discuss in Chapters 4, 5, and 6 are all polynomial, like Karmarkar's method. Further historical information on the development of interior-point methods is given in Chapter 2.

Exercises

1. Explain why we cannot have $\mu_k = 0$ for an iterate $(x^k, \lambda^k, s^k) \in \mathcal{F}^o$ from Framework PD.

2. (i) Show that $\mathcal{N}_2(\theta_1) \subset \mathcal{N}_2(\theta_2)$ when $0 \leq \theta_1 < \theta_2 < 1$ and that $\mathcal{N}_{-\infty}(\gamma_1) \subset \mathcal{N}_{-\infty}(\gamma_2)$ for $0 < \gamma_2 \leq \gamma_1 \leq 1$.

   (ii) Show that $\mathcal{N}_2(\theta) \subset \mathcal{N}_{-\infty}(\gamma)$ if $\gamma \leq 1 - \theta$.

3. Why is the neighborhood $\mathcal{N}_{-\infty}$ called the "one-sided ∞-norm neighborhood"?

4. Given an arbitrary point $(x, \lambda, s) \in \mathcal{F}^o$, find the range of γ values for which $(x, \lambda, s) \in \mathcal{N}_{-\infty}(\gamma)$. (The range depends on x and s.)

5. For $n = 2$, find a point $(x, s) > 0$ for which the condition

$$\|XSe - \mu e\|_2 \leq \theta\mu$$

is *not* satisfied for any $\theta \in (0, 1)$.

6. Show that Φ_ρ defined by (1.18) has the property (1.17a).

7. Write down the KKT conditions for the convex quadratic program (1.22).

8. By eliminating the vector s from (1.21), express the LCP as a convex quadratic program.

9. Use the fact that (x, s) is strictly positive to show that the coefficient matrix in (1.25a) is symmetric positive semidefinite. Show that this matrix is positive definite if A has full row rank.

Chapter 2

Background: Linear Programming and Interior-Point Methods

The linear programming problem is simple to state and visualize, and it has given rise to much interesting theory during the past fifty years. Researchers in complexity theory, convex analysis, mathematical programming, and numerical analysis have examined it closely from their own points of view. All have contributed to a more complete understanding of the mathematics of the problem and to improvements in the algorithms used to solve it.

In this chapter, we touch on just a few aspects of the theory of linear programming—in particular, the aspects that are relevant to primal-dual methods. We give a few simple results from the duality theory and then direct our attention to two major results:

- the Goldman–Tucker result about existence of strictly complementary solutions;
- existence of the central path defined in (1.9).

For a broader and more fundamental treatment of linear programming, its applications and its relationship to network and integer programming, we refer the reader to one of the excellent texts in the area. Chvátal [21], Luenberger [75], and Murtagh [99] give overviews of general linear programming with an emphasis on the simplex method; Ahuja, Magnanti, and Orlin [2] describe the state of the art in network optimization; and Nemhauser and

Wolsey [100] provide the standard reference for integer programming and combinatorial optimization.

The final pages of this chapter contain some background material on interior-point methods. We discuss the logarithmic barrier function and the logarithmic potential functions used by Karmarkar [57] and others. We briefly discuss the primal path-following algorithm of Renegar [112] and the affine-scaling algorithm of Dikin [26]. Turning to primal-dual algorithms, we outline the origins of the central path and the historical development of the various classes of algorithms, including short- and long-step path-following, potential-reduction, and affine-scaling algorithms.

Standard Form

Recall from Chapter 1 that the standard form of the linear programming problem is

$$\min c^T x \text{ subject to } Ax = b,\ x \geq 0, \tag{2.1}$$

where

$$c, x \in \mathbb{R}^n, \qquad b \in \mathbb{R}^m, \qquad A \in \mathbb{R}^{m \times n}.$$

The dual of (2.1) is

$$\max b^T \lambda \text{ subject to } A^T \lambda + s = c,\ s \geq 0, \tag{2.2}$$

where $\lambda \in \mathbb{R}^m$ and $s \in \mathbb{R}^n$.

Simple devices can be used to transform any linear program to the standard form (2.1). For instance, the constraints $Cx \leq d$ can be transformed by introducing a slack vector z to obtain

$$Cx + z = d, \qquad z \geq 0,$$

which is closer to the standard form. Bound constraints of the form $x_i \geq l_i$ (where x_i is the ith component of x and l_i is a finite number) can be transformed to positivity constraints by performing the shift $x_i \leftarrow x_i - l_i$. Components x_i that are not explicitly bounded can be split into positive and negative parts as

$$x_i = x_i^+ - x_i^-, \qquad x_i^+ \geq 0,\ x_i^- \geq 0. \tag{2.3}$$

Linear programming software performs such transformations automatically to convert general linear programs into the standard primal form (2.1), the dual form (2.2), or whatever other form it chooses. Because of the equivalence among the different formulations, nothing is lost in the way of generality by considering just the form (2.1), as we do in most of this book. Computationally, though, some formulations may be more efficient than others for specific problems and algorithms.

Optimality Conditions, Duality, and Solution Sets

In Chapter 1, we noted that the vector (x^*, λ^*, s^*) is a primal-dual solution for (2.1), (2.2) if and only if it satisfies the KKT conditions (see Theorem A.1). We restate the KKT conditions here for convenience as

$$\begin{aligned} A^T\lambda + s &= c, && \text{(2.4a)} \\ Ax &= b, && \text{(2.4b)} \\ x_i s_i &= 0, \quad i = 1, 2, \ldots, n, && \text{(2.4c)} \\ (x, s) &\geq 0. && \text{(2.4d)} \end{aligned}$$

An immediate consequence of the complementarity condition (2.4c) is that $(x^*)^T s^* = 0$ if x^* and s^* satisfy (2.4c). Almost as immediate is the proof of our claim (1.3) from Chapter 1: If (x, s, λ) satisfies the three conditions (2.4a), (2.4b), (2.4d), we have

$$0 \leq s^T x = (c - A^T\lambda)^T x = c^T x - \lambda^T (Ax) = c^T x - b^T \lambda, \qquad (2.5)$$

and hence $b^T\lambda \leq c^T x$. If we have a point (x^*, λ^*, s^*) that satisfies all four conditions in (2.4), it follows from $(x^*)^T s^* = 0$ that $c^T x^* = b^T \lambda^*$, so the optimal values for the primal and dual problems are the same. We denote the common optimal value by Z^*; that is,

$$b^T \lambda^* = Z^* = c^T x^*. \qquad (2.6)$$

Since $c^T x \geq c^T x^*$ for any primal feasible point x and $b^T\lambda \leq b^T\lambda^*$ for any dual feasible point (λ, s), we can write

$$(x, \lambda, s) \in \mathcal{F} \;\; \Rightarrow \;\; b^T\lambda \leq Z^* \leq c^T x, \qquad (2.7)$$

where $\mathcal{F}$ is the primal-dual feasible set defined in (1.7a).

Let us step back a little to see how the primal and dual linear programs are related through the nonlinear system (2.4).

The KKT result, Theorem A.1, gives us a *necessary* condition for optimality: If x is a solution of the primal problem (2.1), there exists a vector (λ, s) such that (2.4) is satisfied. This result also can be proven directly, without recourse to the general KKT theorem. A proof based on Farkas's lemma is sketched in the exercises at the end of this chapter.

Theorem A.2 shows that the KKT conditions are also *sufficient*: If we have a primal feasible vector x, and another vector (λ, s) such that the conditions (2.4) are satisfied, x is a solution of (2.1). We can prove this claim directly by taking an arbitrary primal feasible vector $\bar{x}$ and showing that its objective value is no smaller than $c^T x$. From (2.4), we have

$$c^T \bar{x} = (A^T \lambda + s)^T \bar{x} = b^T \lambda + s^T \bar{x} \geq b^T \lambda = c^T x, \tag{2.8}$$

where we have used the inequality $s^T \bar{x} \geq 0$, which follows from $s \geq 0$ and $\bar{x} \geq 0$. Hence $c^T \bar{x} \geq c^T x$, so x achieves the best possible objective value among all feasible points.

We conclude that the KKT conditions (2.4) are both necessary and sufficient for optimality in the primal problem (2.1). By a similar argument (see the exercises), we can show that the conditions (2.4) are also necessary and sufficient for dual optimality. The primal and dual linear programs (2.1) and (2.2) are flip sides of the same coin.

In the remainder of the book, we use Ω_P and Ω_D to denote the primal and dual solution sets, respectively:

$$\Omega_P = \{x^* \mid x^* \text{ solves } (2.1)\}, \qquad \Omega_D = \{(\lambda^*, s^*) \mid (\lambda^*, s^*) \text{ solves } (2.2)\}.$$

The primal-dual solution set Ω is just the Cartesian product:

$$\Omega = \Omega_P \times \Omega_D = \{(x^*, \lambda^*, s^*) \text{ satisfying } (2.4)\}.$$

The solution sets Ω_P, Ω_D, and Ω all are closed. This claim is easily verified in the case of Ω_P if we use the following equivalent characterization:

$$\Omega_P = \{x^* \mid Ax^* = b, \;\; x^* \geq 0, \;\; c^T x^* = Z^*\}, \tag{2.9}$$

where Z^* is the optimal objective value defined in (2.6). The definition (2.9) shows that Ω_P is an intersection of closed subsets of $\mathbb{R}^n$ and therefore is itself closed. Similar logic applies to Ω_D and Ω.

More on Duality

We prove three elementary results to elaborate on the relationship between the primal and dual problems and to lay the foundation for the primal-dual methods of this book. The first result concerns existence of a primal-dual solution that, as we show, is guaranteed by existence of a feasible primal-dual point.

Theorem 2.1

i. *If the primal and dual problems are both feasible (that is, $\mathcal{F} \neq \emptyset$), the set Ω of primal-dual solutions is nonempty.*

ii. *If either problem (2.1) or (2.2) has an optimal solution, so does the other, and the objective values are equal.*

Proof. Part i. can be proven by applying Corollary 2.7, the variant of Farkas' lemma, to the following system, whose solutions coincide with the primal-dual solution set Ω:

$$\begin{bmatrix} -A & 0 & 0 \\ 0 & I & 0 \\ c^T & 0 & 1 \end{bmatrix} \begin{bmatrix} x \\ s \\ \beta \end{bmatrix} + \begin{bmatrix} 0 \\ A^T \\ -b^T \end{bmatrix} \lambda = \begin{bmatrix} -b \\ c \\ 0 \end{bmatrix}, \qquad (x, s, \beta) \geq 0.$$

We leave details to the reader.

For ii., suppose that (2.1) has an optimal solution x^*. Then, since the KKT conditions (2.4) are *necessary*, there exist vectors λ^* and s^* such that (x^*, λ^*, s^*) satisfies the KKT conditions (2.4). From our comments in Chapter 1, (x^*, λ^*, s^*) is a primal-dual solution; hence, in particular, (λ^*, s^*) is a dual solution. Equality of the optimal objective values follows from (2.5), as noted earlier.

A similar argument can be made if we start by assuming that the dual problem (2.2) has a solution. □

It can happen that the primal or dual linear program (possibly both) is infeasible. When exactly one of the two problems is infeasible, Theorem 2.1 tells us something about the other problem, as we explain in the following result.

Corollary 2.2 *Suppose that the primal problem* (2.1) *is feasible. Then its objective function c^Tx is bounded below on its feasible region if and only if the dual problem* (2.2) *is feasible.*

Similarly, suppose that the dual problem is feasible. Then its objective function $b^T\lambda$ is bounded above on its feasible region if and only if the primal problem (2.1) *is feasible.*

Proof. We prove only the first statement in the theorem, since the proof of the second is identical.

Suppose that the primal problem is feasible. If the dual is also feasible, then by Theorem 2.1i., the primal solution set Ω_P is nonempty and c^Tx is bounded below on the feasible region by c^Tx^* for any $x^* \in \Omega_P$.

For the converse, suppose that the dual is infeasible. Then there is *no* vector (λ, s) such that

$$s + A^T\lambda = c, \qquad s \geq 0.$$

By applying Corollary 2.7 to this system, we deduce that there is a vector d such that

$$d \geq 0, \qquad Ad = 0, \qquad c^T d < 0.$$

Hence, if $\hat{x}$ is any primal feasible point, the ray defined by

$$\{\hat{x} + \alpha d \mid \alpha \geq 0\}$$

lies entirely within the primal feasible set and $c^T x$ is unbounded below along this ray. □

Theorem 2.1 and Corollary 2.2 indicate that the properties of the primal solution set Ω_P are closely related to the existence of feasible points for the dual problem (and similarly for Ω_D and the primal feasible set). We elaborate on this relationship in the next result, since it is important to the analysis of interior-point methods, but we first need another definition.

A vector x is said to be *strictly feasible* for the primal problem if

$$Ax = b, \qquad x > 0;$$

that is, the positivity condition $x \geq 0$ is replaced by a *strict* inequality. Similarly, the vector (λ, s) is strictly feasible for the dual problem if

$$A^T\lambda + s = c, \qquad s > 0.$$

The next result reveals a condition for existence and boundedness of the solution sets Ω_P and Ω_D.

Theorem 2.3 *Suppose that the primal and dual problems are feasible; that is, $\mathcal{F} \neq \emptyset$. Then if the dual problem has a strictly feasible point, the primal solution set Ω_P is nonempty and bounded. Similarly, if the primal problem has a strictly feasible point, the set*

$$\{s^* \mid (\lambda^*, s^*) \in \Omega_D \text{ for some } \lambda^* \in \mathbb{R}^m\}$$

is nonempty and bounded.

Proof. We prove the first statement and leave the second as an exercise.

Denote the strictly feasible dual point by $(\bar{\lambda}, \bar{s})$, and let $\hat{x}$ be any primal feasible point (not necessarily strict). Then as in (2.5) we have

$$0 \le \bar{s}^T \hat{x} = c^T \hat{x} - b^T \bar{\lambda}. \tag{2.10}$$

Now consider the set of points $\mathcal{T}$ defined by

$$\mathcal{T} = \{x \mid Ax = b, \quad x \ge 0, \quad c^T x \le c^T \hat{x}\}.$$

The set $\mathcal{T}$ is nonempty (since $\hat{x} \in \mathcal{T}$) and is obviously closed. For any $x \in \mathcal{T}$, we have from (2.5) and (2.10) that

$$\sum_{i=1}^{n} \bar{s}_i x_i = \bar{s}^T x = c^T x - b^T \bar{\lambda} \le c^T \hat{x} - b^T \bar{\lambda} = \bar{s}^T \hat{x}.$$

Since all terms in the summation on the left are nonnegative, we have

$$x_i \le \frac{1}{\bar{s}_i} \bar{s}^T \hat{x} \quad \Rightarrow \quad \|x\|_\infty \le \left(\max_{i=1,2,\ldots,n} \frac{1}{\bar{s}_i} \right) \bar{s}^T \hat{x}.$$

Since x was an arbitrary element of $\mathcal{T}$, we conclude from this inequality that $\mathcal{T}$ is bounded, as well as nonempty and closed. Hence, the function $c^T x$ must attain its minimum value on $\mathcal{T}$. That is, there exists a point x^* such that

$$x^* \in \mathcal{T}, \qquad c^T x^* \le c^T x \text{ for all } x \in \mathcal{T}.$$

Obviously, the set of points x^* with this property coincides with Ω_P. Hence, Ω_P is nonempty and bounded, since it is a subset of the bounded set $\mathcal{T}$. □

The converse of Theorem 2.3 is also true: If the primal solution set Ω_P is nonempty and bounded, the dual problem has a strictly feasible point (similarly for Ω_D and the primal feasible set). The proof makes use of a "theorem of the alternative" (see, for example, Mangasarian [81, Chapter 2]), and we omit it here.

The $\mathcal{B} \cup \mathcal{N}$ Partition and Strict Complementarity

For every solution (x^*, λ^*, s^*), we know from (2.4c) that

$$x_j^* = 0 \text{ and/or } s_j^* = 0 \text{ for all } j = 1, 2, \ldots, n.$$

We can define two index sets $\mathcal{B}$ and $\mathcal{N}$ as follows.

$$\mathcal{B} = \{j \in \{1, 2, \ldots, n\} \mid x_j^* \ne 0 \text{ for some } x^* \in \Omega_P\}, \tag{2.11a}$$
$$\mathcal{N} = \{j \in \{1, 2, \ldots, n\} \mid s_j^* \ne 0 \text{ for some } (\lambda^*, s^*) \in \Omega_D\}. \tag{2.11b}$$

In fact, $\mathcal{B}$ and $\mathcal{N}$ form a partition of the index set $\{1, 2, \dots, n\}$; every $j = 1, 2, \dots, n$ belongs to either $\mathcal{B}$ or $\mathcal{N}$ but not both. It is easy enough to prove half of this result, namely, that $\mathcal{B}$ and $\mathcal{N}$ are disjoint. If there happened to be an index j that belonged to both sets, then from (2.11) there would be a primal solution x^* and a dual solution (λ^*, s^*) such that $x_j^* > 0$ and $s_j^* > 0$. But then we would have $x_j^* s_j^* > 0$, contradicting the complementarity condition (2.4c). Hence, $\mathcal{B} \cap \mathcal{N} = \emptyset$.

The result $\mathcal{B} \cup \mathcal{N} = \{1, 2, \dots, n\}$ is known as the Goldman–Tucker theorem [41]. Since this theorem is crucial to the local and superlinear convergence analysis of primal-dual methods, we state it formally and give a proof at the end of the chapter.

Theorem 2.4 (Goldman–Tucker) $\mathcal{B} \cup \mathcal{N} = \{1, 2, \dots, n\}$. *That is, there exist at least one primal solution $x^* \in \Omega_P$ and one dual solution $(\lambda^*, s^*) \in \Omega_D$ such that $x^* + s^* > 0$.*

Primal-dual solutions (x^*, λ^*, s^*) with the property $x^* + s^* > 0$ are known as *strictly complementary* solutions. Theorem 2.4 guarantees that at least one such solution exists. A given linear program may have multiple primal-dual solutions, some that are strictly complementary and others that are not, as the following trivial example illustrates. Consider

$$\min_{x \in \mathbb{R}^3} x_1 \text{ subject to } x_1 + x_2 + x_3 = 1, \; x \geq 0, \tag{2.12}$$

whose dual is

$$\max_{\lambda \in \mathbb{R},\, s \in \mathbb{R}^3} \lambda \text{ subject to } \begin{bmatrix} 1 \\ 1 \\ 1 \end{bmatrix} \lambda + s = \begin{bmatrix} 1 \\ 0 \\ 0 \end{bmatrix}, \quad s \geq 0. \tag{2.13}$$

Primal-dual solutions for (2.12) and (2.13) are

$$x^* = (0, t, 1-t)^T, \quad \lambda^* = 0, \quad s^* = (1, 0, 0)^T \quad \text{for any } t \in [0, 1]. \tag{2.14}$$

For t in the *open* interval $(0, 1)$, it is clear that (x^*, λ^*, s^*) is a strictly complementary solution. If t takes on one of its extreme values 0 or 1, however, the primal-dual solution is no longer strictly complementary, since there is an index j for which x_j^* and s_j^* are both zero.

It is easy to show that there is a primal solution x^* for which $x_j^* > 0$ for *all* $j \in \mathcal{B}$; that is, $x_{\mathcal{B}}^* > 0$. By the definition (2.11a), there exist solutions

$\bar{x}^j \in \Omega_P$ for each $j \in \mathcal{B}$ with the property that $\bar{x}^j_j > 0$. The average of these solutions, defined by

$$x^* = \frac{1}{|\mathcal{B}|} \sum_{j \in \mathcal{B}} \bar{x}^j, \tag{2.15}$$

is feasible, has the same objective value as each $\bar{x}^j$, and hence is also a solution. It is easy to check that this x^* has the desired property $x^*_{\mathcal{B}} > 0$.

Similarly, we can show that there is a dual solution (λ^*, s^*) such that $s^*_{\mathcal{N}} > 0$.

A Strictly Interior Point

The set of strictly feasible primal-dual vectors $\mathcal{F}^o$ was defined in (1.7b) by changing the nonnegativity conditions $(x, s) \geq 0$ into positivity conditions $(x, s) > 0$. The primal-dual algorithms discussed in Chapters 4 and 5 keep all their iterates (x^k, λ^k, s^k) inside this set. As they reduce the duality gap $x^T s$ to zero, some of the components of x and s will inevitably go to zero, so the iterates approach the boundary of $\mathcal{F}^o$ without ever actually leaving this set.

However, many linear programming problems have *no* strictly feasible points—that is, $\mathcal{F}^o = \emptyset$—although they still may be feasible ($\mathcal{F} \neq \emptyset$) and still may have finite optimal solutions. Consider the simple problem

$$\min_{x \in \mathbb{R}^3} x_1 \text{ subject to } x_1 + x_3 = 0, \; x \geq 0,$$

and its dual

$$\max_{\lambda \in \mathbb{R}, s \in \mathbb{R}^3} 0 \text{ subject to } \begin{bmatrix} 1 \\ 0 \\ 1 \end{bmatrix} \lambda + \begin{bmatrix} s_1 \\ s_2 \\ s_3 \end{bmatrix} = \begin{bmatrix} 1 \\ 0 \\ 0 \end{bmatrix}, \quad s \geq 0.$$

Any feasible primal-dual vector $(x, \lambda, s) \in \mathcal{F}$ has $x_1 = x_3 = s_2 = 0$, so $\mathcal{F}^o = \emptyset$. The optimal objective function value for this problem is 0, and the primal-dual solution set is defined by

$$x^* = \begin{bmatrix} 0 \\ x_2^* \\ 0 \end{bmatrix}, \qquad s^* = \begin{bmatrix} 1 - \lambda^* \\ 0 \\ -\lambda^* \end{bmatrix},$$

where x_2^* and λ^* are any numbers for which $x_2^* \geq 0$ and $\lambda^* \leq 0$.

Although many primal-dual algorithms require a strictly feasible starting point, they can be adapted to handle the case in which no such point exists.

One way of dealing with $\mathcal{F}^o = \emptyset$ is to generalize the algorithms so that they accept infeasible starting points; we take this tack in Chapter 6. A second approach is to embed the problem into a slightly larger problem that *does* have strictly feasible points and apply the standard algorithms to the reformulated problem. This approach is described in Chapter 9.

An immediate consequence of Theorem 2.3 is that when a strictly feasible primal-dual point exists, the x^* and s^* components of all primal-dual solutions lie in a bounded set; that is,

$$\mathcal{F}^o \neq \emptyset \;\Rightarrow\; \{(x^*, s^*) \,|\, (x^*, \lambda^*, s^*) \in \Omega \text{ for some } \lambda^*\} \quad \text{is a bounded set.} \tag{2.16}$$

This assertion also follows from the following result, which is used later in the chapter in the proof of existence of the central path $\mathcal{C}$.

Lemma 2.5 *Suppose that $\mathcal{F}^o \neq \emptyset$. Then for each $K \geq 0$, the set*

$$\{(x, s) \,|\, (x, \lambda, s) \in \mathcal{F} \text{ for some } \lambda, \text{ and } x^T s \leq K\}$$

is bounded.

Proof. Let $(\bar{x}, \bar{\lambda}, \bar{s})$ be any vector in $\mathcal{F}^o$ and (x, λ, s) be any point in $\mathcal{F}$ with $x^T s \leq K$. Since $A\bar{x} = b$ and $Ax = b$, we have $A(\bar{x} - x) = 0$. Similarly, $A^T(\bar{\lambda} - \lambda) + (\bar{s} - s) = 0$. These two equations imply that

$$(\bar{x} - x)^T(\bar{s} - s) = -(\bar{x} - x)^T A^T(\bar{\lambda} - \lambda) = 0.$$

Rearranging this system and using $x^T s \leq K$, we find that

$$\bar{x}^T s + \bar{s}^T x \leq K + \bar{x}^T \bar{s}. \tag{2.17}$$

Since $(\bar{x}, \bar{s}) > 0$, the quantity ξ defined by

$$\xi = \min_{i=1,2,\ldots,n} \min(\bar{x}_i,\, \bar{s}_i)$$

is positive. Substituting in (2.17), we obtain

$$\xi e^T(x + s) \leq K + \bar{x}^T \bar{s},$$

which implies that

$$0 \leq x_i \leq \frac{1}{\xi}(K + \bar{x}^T \bar{s}), \qquad 0 \leq s_i \leq \frac{1}{\xi}(K + \bar{x}^T \bar{s}), \qquad i = 1, 2, \ldots, n,$$

proving the result. □

When it is nonempty, the strictly feasible set $\mathcal{F}^o$ coincides with the relative interior ri($\mathcal{F}$) of the feasible set $\mathcal{F}$. See Rockafellar [116] for a definition of the relative interior and a description of the important role of this concept in convex analysis.

Rank of the Matrix A

When discussing algorithms for linear programming, it is often convenient to assume that the matrix A in (2.1) has full row rank m. There exist perfectly valid linear programs that do not satisfy this assumption, but such problems can always be transformed to equivalent problems that *do* satisfy it. For instance, the trivial problem

$$\min x_1 + x_2 \quad \text{subject to} \quad \begin{bmatrix} 1 & 2 \\ -2 & -4 \end{bmatrix} \begin{bmatrix} x_1 \\ x_2 \end{bmatrix} = \begin{bmatrix} 3 \\ -6 \end{bmatrix}, \quad \begin{bmatrix} x_1 \\ x_2 \end{bmatrix} \geq 0,$$

has the (primal) solution $x^* = (0, 3/2)$, but the 2×2 constraint matrix is rank deficient. The second constraint is obviously redundant, so we can eliminate it and restate the problem equivalently as

$$\min x_1 + x_2 \quad \text{subject to} \quad \begin{bmatrix} 1 & 2 \end{bmatrix} \begin{bmatrix} x_1 \\ x_2 \end{bmatrix} = 3, \quad \begin{bmatrix} x_1 \\ x_2 \end{bmatrix} \geq 0.$$

In general, the matrix A can be reduced to a full rank matrix via QR factorization or Gaussian elimination with column pivoting. A QR factorization applied to the consistent system $Ax = b$ obtains an $m \times m$ orthogonal matrix Q such that

$$QA = \begin{bmatrix} \bar{A} \\ 0 \end{bmatrix}, \qquad Qb = \begin{bmatrix} \bar{b} \\ 0 \end{bmatrix},$$

where $\bar{A}$ and $\bar{b}$ have the same number of rows and $\bar{A}$ has full row rank. The systems $Ax = b$ and $\bar{A}x = \bar{b}$ are equivalent; that is, any vector x that satisfies one of these equations also satisfies the other. Details of the QR factorization are given in Appendix A.

Implementations of the simplex algorithm usually introduce artificial variables into the formulation during the determination of an initial basis ("Phase I"). In this augmented formulation, the constraint matrix A has full rank, so the issue of rank deficiency is not of great concern in practice.

Interior-point codes usually perform no such augmentation, but rank deficiency is sometimes eliminated during the presolve (see Chapter 11). Even when it is not, these codes usually produce satisfactory solutions to the problem, with the help of some special features in the linear algebra routines.

The linear systems of equations (1.12) and (1.20) that define primal-dual steps have unique solutions in their Δx and Δs components even if A is rank deficient. To verify this claim, we show that all solutions $(\Delta x, \Delta\lambda, \Delta s)$ of the following homogeneous system must have $\Delta x = 0$ and $\Delta s = 0$:

$$\begin{bmatrix} 0 & A^T & I \\ A & 0 & 0 \\ S & 0 & X \end{bmatrix} \begin{bmatrix} \Delta x \\ \Delta\lambda \\ \Delta s \end{bmatrix} = \begin{bmatrix} 0 \\ 0 \\ 0 \end{bmatrix}, \tag{2.18}$$

where X and S are defined by (1.6), with $(x, s) > 0$. From the first two block rows of this system, we have

$$\Delta x^T \Delta s = -\Delta x^T A^T \Delta\lambda = -(A\Delta x)^T \Delta\lambda = 0.$$

Let $D = S^{-1/2}X^{1/2}$ be the diagonal matrix defined by (1.24), whose diagonal elements are strictly positive. If we multiply the last block row in (2.18) by $(XS)^{-1/2}$, we obtain

$$D^{-1}\Delta x + D\Delta s = 0.$$

Using $\Delta x^T \Delta s = 0$, we obtain

$$\begin{aligned} 0 &= \|D^{-1}\Delta x + D\Delta s\|_2^2 \\ &= \|D^{-1}\Delta x\|_2^2 + 2\Delta x^T\Delta s + \|D\Delta s\|_2^2 = \|D^{-1}\Delta x\|_2^2 + \|D\Delta s\|_2^2. \end{aligned}$$

Hence $D^{-1}\Delta x = 0$ and $D\Delta s = 0$, which implies $\Delta x = 0$ and $\Delta s = 0$, as claimed.

If A has full rank, the $\Delta\lambda$ component is also unique. This claim also follows from (2.18): If we substitute $\Delta s = 0$ into the first block row, we obtain $A^T\Delta\lambda = 0$ and therefore $\Delta\lambda = 0$ by the full rank property.

Bases and Vertices

The set $\mathcal{B}$ defined above does not necessarily contain m elements, where m is the number of rows in A. Any number of elements from 0 to n is possible. In the example (2.12), (2.13), for instance, we have $m = 1$ and yet $|\mathcal{B}| = 2$. This point sometimes confuses people who are familiar with the simplex method, in which an index set known as the *basis* plays a major role. Each basis $\mathcal{M}$ contains exactly m indices, chosen so that

- the $m \times m$ column submatrix of A that corresponds to the basis $\mathcal{M}$ is nonsingular; that is,

$$B = [A_{\cdot i}]_{i \in \mathcal{M}} \qquad \text{is nonsingular.} \tag{2.19}$$

- there is a vector $x \in \mathbb{R}^n$ that satisfies

$$Ax = b, \quad x \geq 0, \quad x_i = 0 \text{ for } i \notin \mathcal{M}. \tag{2.20}$$

The vector x in (2.20) is unique for each given basis $\mathcal{M}$ and is known as a *basic feasible point.* By combining (2.20) and (2.19), we can see that basic feasible points have the form

$$x = \Pi \begin{bmatrix} x_B \\ 0 \end{bmatrix} = \Pi \begin{bmatrix} B^{-1}b \\ 0 \end{bmatrix}, \tag{2.21}$$

where Π is some $n \times n$ permutation matrix.

Because of the correspondence (2.20) between bases $\mathcal{M}$ and basic feasible points x, we can think of the simplex algorithm as generating a sequence of *bases* rather than a sequence of vectors x^k (Dantzig [24]). At each simplex iteration, one index leaves the current basis and another enters, and x changes accordingly. When the linear program has a solution, the simplex method eventually finds an *optimal basis* and a corresponding primal solution vector x^*. In the example (2.12), there are three basic feasible points $x^* = (1,0,0)^T$, $x^* = (0,1,0)^T$, and $x^* = (0,0,1)^T$, corresponding to the bases $\mathcal{M} = \{1\}$, $\mathcal{M} = \{2\}$, and $\mathcal{M} = \{3\}$, respectively.

This example suggests a geometric significance for basic feasible points. The feasible set for (2.1) is a polygon, consisting of the intersection of the plane defined by $Ax = b$ with the nonnegative orthant defined by $x \geq 0$. The *vertices* of this polygon are its extreme points, the points that do not lie on a straight line connecting two other points in the polygon. Vertices are easily recognizable; see Figure 2.1. We can show that all basic feasible points are vertices of the feasible polygon $\{x \mid Ax = b, \ x \geq 0\}$, and vice versa. A proof of this claim, and the other claims above, can be found in Nocedal and Wright [104].

Since all iterates of the simplex algorithm are vertices, it follows that this algorithm will terminate at a vertex solution, even if an entire edge or face of the feasible polygon is optimal. The basis $\mathcal{M}$ is an optimal basis if it corresponds to an optimal vertex.

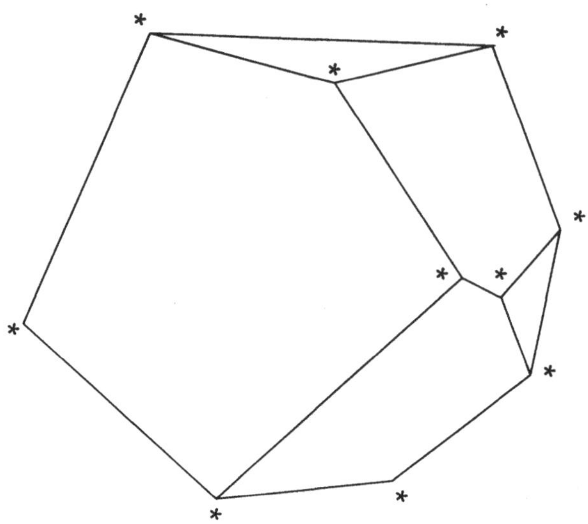

Figure 2.1. Vertices (indicated by $*$) of a three-dimensional polygon.

Interior-point methods tend, in the case of nonunique solutions, to approach points in the relative interior of the optimal set, rather than vertices. However, vertices/basic feasible points play a role in the proof of polynomial complexity of interior-point methods, as we see in Chapter 3.

Farkas's Lemma and a Proof of the Goldman–Tucker Result

Farkas's lemma is a famous result in mathematical programming that can be stated as follows.

Lemma 2.6 (Farkas's lemma) *For each matrix $G \in \mathbb{R}^{p \times n}$ and each vector $g \in \mathbb{R}^n$, either*

I. $Gd \geq 0,\ g^T d < 0,$ *has a solution* $d \in \mathbb{R}^n$,

II. $G^T \pi = g,\ \pi \geq 0,$ *has a solution* $\pi \in \mathbb{R}^p$,

but never both.

Geometrically, the result says that if we define a cone by taking a convex combination of the rows of G, then either g lies within the cone (case II) or there is a plane defined by d that separates g from the cone (case I).

We obtain a simple corollary of Farkas's lemma by introducing some equalities.

Corollary 2.7 *For each pair of matrices $G \in \mathbb{R}^{p \times n}$ and $H \in \mathbb{R}^{q \times n}$ and each vector $g \in \mathbb{R}^n$, either*

I. $Gd \geq 0$, $Hd = 0$, $g^T d < 0$, *has a solution* $d \in \mathbb{R}^n$,

II. $G^T \pi + H^T \eta = g$, $\pi \geq 0$, *has a solution* $\pi \in \mathbb{R}^p$, $\eta \in \mathbb{R}^q$,

but never both.

This corollary can be used to prove the Goldman–Tucker result.

Proof. (Theorem 2.4) Let $\mathcal{J}$ be the set of indices in $\{1, 2, \ldots, n\}$ that do not belong to either $\mathcal{B}$ or $\mathcal{N}$. We prove the result by showing that $\mathcal{J}$ is empty.

We have already proved that $\mathcal{B} \cap \mathcal{N} = \emptyset$; hence, $\mathcal{B} \cup \mathcal{N} \cup \mathcal{J}$ is a partition of $\{1, 2, \ldots, n\}$. Let $A_{\mathcal{B}}$ and $A_{\mathcal{J}}$ denote the submatrices of columns of A that correspond to $\mathcal{B}$ and $\mathcal{J}$, respectively.

Select any index $i \in \mathcal{J}$. We show that i must also belong to either $\mathcal{N}$ or $\mathcal{B}$, depending on whether there exists a vector w satisfying the following properties:

$$\begin{aligned} A_{\cdot i}^T w &< 0, \\ -A_{\cdot j}^T w &\geq 0 \qquad \text{for } j \in \mathcal{J} \backslash \{i\}, \\ A_{\mathcal{B}}^T w &= 0, \end{aligned} \tag{2.22}$$

where $A_{\cdot i}$ denotes the ith column of A. Suppose first that w satisfying (2.22) exists. Let (x^*, λ^*, s^*) be a primal-dual solution for which $s_{\mathcal{N}}^* > 0$, and define the vector $(\bar{\lambda}, \bar{s})$ as

$$\bar{\lambda} = \lambda^* + \epsilon w, \qquad \bar{s} = c - A^T \bar{\lambda} = s^* - \epsilon A^T w,$$

choosing $\epsilon > 0$ small enough that

$$\begin{aligned} \bar{s}_i &= s_i^* - \epsilon A_{\cdot i}^T w > 0, \\ \bar{s}_j &= s_j^* - \epsilon A_{\cdot j}^T w \geq 0, \qquad j \in \mathcal{J} \backslash \{i\}, \\ \bar{s}_{\mathcal{B}} &= s_{\mathcal{B}}^* = 0, \\ \bar{s}_{\mathcal{N}} &= s_{\mathcal{N}}^* - \epsilon A_{\mathcal{N}}^T w > 0. \end{aligned}$$

It follows from these relations that $(\bar{\lambda}, \bar{s})$ is feasible for the dual problem. In fact, it is also optimal, since any primal solution vector x^* must have $x_{\mathcal{N}}^* = 0$ and therefore $\bar{s}^T x^* = 0$. Therefore, by the definition (2.11b), we must have $i \in \mathcal{N}$.

Suppose, alternatively, that no vector w satisfies (2.22). Applying Corollary 2.7, we deduce that the following system must have a solution:

$$-\sum_{j\in\mathcal{J}\backslash\{i\}} \pi_j A_{\cdot j} + A_{\mathcal{B}}\eta = A_{\cdot i}, \qquad \pi_j \geq 0 \text{ for all } j \in \mathcal{J}\backslash\{i\}. \tag{2.23}$$

By defining a vector $v \in \mathbb{R}^{|\mathcal{J}|}$ as

$$v_i = 1, \qquad v_j = \pi_j \quad \text{for all } j \in \mathcal{J}\backslash\{i\},$$

we can rewrite (2.23) as

$$A_{\mathcal{J}}v = A_{\mathcal{B}}\eta, \qquad v \geq 0, \qquad v_i > 0. \tag{2.24}$$

Now let x^* be a primal solution for which $x^*_{\mathcal{B}} > 0$, and define $\bar{x}$ by

$$\bar{x}_{\mathcal{B}} = x^*_{\mathcal{B}} - \epsilon\eta, \qquad \bar{x}_{\mathcal{J}} = \epsilon v, \qquad \bar{x}_{\mathcal{N}} = 0.$$

Substituting from (2.24), we have that $A\bar{x} = b$, and for sufficiently small $\epsilon > 0$, we also have that $\bar{x} \geq 0$. So $\bar{x}$ is feasible and, in fact, optimal because of $\bar{x}_{\mathcal{N}} = 0$. Since $v_i = 1$, we also have that $\bar{x}_i = \epsilon > 0$; hence, $i \in \mathcal{B}$ by (2.11a).

We have shown that any index $i \in \mathcal{J}$ also belongs to either $\mathcal{B}$ or $\mathcal{N}$. Therefore, by the definition of $\mathcal{J}$, we have $\mathcal{J} = \emptyset$; therefore, the proof is complete. □

The Central Path

In Chapter 1, we defined the central path $\mathcal{C}$ to be an arc of points $(x_\tau, \lambda_\tau, s_\tau) \in \mathcal{F}^o$ parametrized by a positive scalar τ. Each point on $\mathcal{C}$ satisfies the following KKT-like system for some $\tau > 0$:

$$A^T\lambda + s = c, \tag{2.25a}$$
$$Ax = b, \tag{2.25b}$$
$$XSe = \tau e, \tag{2.25c}$$
$$(x, s) > 0, \tag{2.25d}$$

where we again have used the notation

$$X = \text{diag}(x_1, x_2, \ldots, x_n), \qquad S = \text{diag}(s_1, s_2, \ldots, s_n), \qquad e = (1, 1, \ldots, 1)^T.$$

The key feature of points on $\mathcal{C}$ is that the pairwise products $x_i s_i$ are identical for all i.

The central path also can be derived from the sequence of minimizers of the log barrier functions for either the primal or dual linear program. Later in this chapter, we define the log barrier function (2.33) for the primal problem (2.1) and show through the KKT conditions for this function that its minimizers x_τ are simply the x components of the central path vectors $(x_\tau, \lambda_\tau, s_\tau)$.

As we mentioned in Chapter 1, the central path stabilizes primal-dual algorithms by providing a route that can be followed to the solution set. We will have much more to say about the role of $\mathcal{C}$ in later chapters. Our concern here is to show that $\mathcal{C}$ actually exists. We prove that if $\mathcal{F}^o$ is nonempty, then (2.25) has a solution $(x_\tau, \lambda_\tau, s_\tau)$ for every $\tau > 0$. Moreover, the x and s components of this solution are unique.

This result is more difficult to obtain than one might expect. Our approach is to define a reduced strictly feasible set $\mathcal{H}^o$ by

$$\mathcal{H}^o = \{(x, s) \mid (x, \lambda, s) \in \mathcal{F}^o \text{ for some } \lambda \in \mathbb{R}^m\}$$

and then to identify (x_τ, s_τ) as the unique minimizer over $\mathcal{H}^o$ of a barrier function $f_\tau(x, s)$ defined by

$$f_\tau(x, s) = \frac{1}{\tau} x^T s - \sum_{j=1}^{n} \log(x_j s_j).$$

Given (x_τ, s_τ), we know by definition of $\mathcal{H}^o$ that there is a $\lambda_\tau \in \mathbb{R}^m$ such that $(x_\tau, \lambda_\tau, s_\tau)$ solves (2.25). Unlike the x and s components, however, λ_τ is not defined uniquely unless A has full row rank.

The function $f_\tau(x, s)$ approaches ∞ whenever (x, s) approaches the boundary of $\mathcal{H}^o$, that is, whenever any of the primal-dual pairs $x_j s_j$ approaches zero. We must therefore have $(x_\tau, s_\tau) > 0$. Three other observations about f_τ are in order:

• *f_τ is strictly convex over $\mathcal{H}^o$.* The first term in f_τ is ostensibly quadratic, but its restriction to $\mathcal{H}^o$ is linear. If $\bar{x}$ is any vector for which $A\bar{x} = b$, we have for any $(x, s) \in \mathcal{H}^o$ that

$$x^T s = c^T x - b^T \lambda = c^T x - \bar{x}^T A^T \lambda = c^T x - \bar{x}^T (c - s) = c^T x + \bar{x}^T s - \bar{x}^T c,$$

which is linear in (x, s). Meanwhile, the summation term in f_τ has a Hessian that is positive definite for all $(x, s) > 0$, so this term is strictly convex on $\mathcal{H}^o$. Hence, the entire function f_τ is strictly convex on $\mathcal{H}^o$.

• *f_τ is bounded below on $\mathcal{H}^o$.* To verify this claim, rewrite f_τ as

$$f_\tau(x,s) = \sum_{j=1}^{n} g\left(\frac{x_j s_j}{\tau}\right) + n - n\log\tau, \tag{2.26}$$

where

$$g(t) = t - \log t - 1.$$

It is easy to show that

$g(t)$ is strictly convex on $(0,\infty)$;

$g(t) \geq 0$ for $t \in (0,\infty)$, with equality if and only if $t = 1$;

$g(t) \uparrow \infty$ as $t \to 0$ or $t \to \infty$.

Using $g(t) \geq 0$ in (2.26), we have

$$f_\tau(x,s) \geq n(1 - \log\tau), \tag{2.27}$$

as claimed.

• *Given $\tau > 0$, and any number κ, all points (x,s) in the level set $\{(x,s) \in \mathcal{H}^o \mid f_\tau(x,s) \leq \kappa\}$ satisfy*

$$x_i \in [M_l, M_u], \qquad s_i \in [M_l, M_u], \qquad i = 1,2,\ldots,n, \tag{2.28}$$

for some positive numbers M_l and M_u. Because of (2.26), we have $f_\tau(x,s) \leq \kappa$ if and only if

$$\sum_{j=1}^{n} g\left(\frac{x_j s_j}{\tau}\right) \leq \bar{\kappa},$$

where $\bar{\kappa} = \kappa - n + n\log\tau$. Choosing a particular index i, we have

$$g\left(\frac{x_i s_i}{\tau}\right) \leq \bar{\kappa} - \sum_{j\neq i} g\left(\frac{x_j s_j}{\tau}\right) \leq \bar{\kappa},$$

where the last inequality follows from $g(t) \geq 0$. Because of the third listed property of $g(\cdot)$ above, there must be a number M such that

$$\frac{1}{M} \leq x_i s_i \leq M, \qquad i = 1,2,\ldots,n. \tag{2.29}$$

Summing the terms in this expression, we obtain

$$x^T s = \sum_{i=1}^{n} x_i s_i \le nM. \tag{2.30}$$

Because of our earlier boundedness result—Lemma 2.5—we have from (2.30) that there is a number M_u such that $x_i \in (0, M_u]$ and $s_i \in (0, M_u]$ for all $i = 1, 2, \ldots, n$. Now, using (2.29), we have $x_i \ge 1/(Ms_i) \ge 1/(MM_u)$ for all i (and similarly for s_i). The claim (2.28) holds if we set $M_l = 1/(MM_u)$.

We are now in a position to show that for any $\tau > 0$, f_τ achieves its minimum value in $\mathcal{H}^o$, that this minimizer is unique, and that it can be used to construct a solution to (2.25).

Theorem 2.8 *Suppose that $\mathcal{F}^o \neq \emptyset$, and let τ be any positive number. Then f_τ has a unique local minimizer in $\mathcal{H}^o$, and the central path conditions* (2.25) *have a solution.*

Proof. We showed above that each level set of f_τ is contained in a compact subset of its domain $\mathcal{H}^o$, so f_τ attains its minimum value in $\mathcal{H}^o$. Since f_τ is strictly convex in $\mathcal{H}^o$, this solution is unique (see Theorem A.2 in Appendix A).

We now show that the minimizer of f_τ makes up the x and s components of the solution of (2.25). This minimizer solves the problem

$$\min f_\tau(x, s) \text{ subject to } Ax = b,\ A^T\lambda + s = c,\ (x, s) > 0. \tag{2.31}$$

Applying the KKT conditions for linearly constrained optimization (Theorem A.1), we find that there are Lagrange multiplier vectors v and w for which the following conditions hold:

$$\frac{\partial}{\partial x} f_\tau(x, s) = A^T v \quad \Rightarrow \quad \frac{s}{\tau} - X^{-1}e = A^T v, \tag{2.32a}$$

$$\frac{\partial}{\partial \lambda} f_\tau(x, s) = Aw \quad \Rightarrow \quad 0 = -Aw, \tag{2.32b}$$

$$\frac{\partial}{\partial s} f_\tau(x, s) = w \quad \Rightarrow \quad \frac{x}{\tau} - S^{-1}e = w, \tag{2.32c}$$

where X, S, and e are defined after (2.25). By combining (2.32b) and (2.32c), we find that

$$A\left(\frac{x}{\tau} - S^{-1}e\right) = 0.$$

Now, taking the inner product of $(x/\tau - S^{-1}e)$ and (2.32a), we obtain

$$\left(\frac{1}{\tau}Xe - S^{-1}e\right)^T \left(\frac{1}{\tau}Se - X^{-1}e\right) = 0.$$

Therefore,

$$\begin{aligned} 0 &= \left(\frac{1}{\tau}Xe - S^{-1}e\right)^T (X^{-1/2}S^{1/2})(X^{1/2}S^{-1/2}) \left(\frac{1}{\tau}Se - X^{-1}e\right) \\ &= \left\|\frac{1}{\tau}(XS)^{1/2}e - (XS)^{-1/2}e\right\|^2. \end{aligned}$$

Hence, we have $\frac{1}{\tau}(XS)^{1/2}e - (XS)^{-1/2}e = 0$, and therefore $XSe = \tau e$.

We have shown that the minimizer (x, s) of f_τ together with the λ component from the feasibility conditions (2.31) satisfies the conditions (2.25), so the proof is complete. □

Background: Primal Methods

Much of the historical motivation for primal-dual methods came from methods such as Karmarkar's that focused solely on the primal problem (2.1). We briefly review some of the landmarks in the development of primal methods that contributed to concepts such as the central path that later became important in the primal-dual context. Primal algorithms are interesting in their own right, too, not least because current research into interior-point algorithms for nonlinear programming incorporates ideas from primal algorithms as well as primal-dual algorithms.

An important element in some primal algorithms is the logarithmic barrier function approximation to (2.1), which is defined as

$$\min_x \; c^T x - \tau \sum_{i=1}^{n} \log x_i \quad \text{subject to } Ax = b,\; x > 0, \tag{2.33}$$

where $\tau > 0$ is a positive parameter. Note that the domain of this function is the set of strictly feasible points for the problem (2.1). By Theorem A.1, the solution x_τ of (2.33) for each value $\tau > 0$ satisfies the conditions

$$\begin{aligned} \tau X^{-1}e + A^T\lambda &= c, && (2.34a) \\ Ax &= b, && (2.34b) \\ x &> 0, && (2.34c) \end{aligned}$$

for some $\lambda \in \mathbb{R}^m$, where X is defined by (1.6). If we define the vector s by

$$s_i = \frac{\tau}{x_i}, \qquad i = 1, 2, \ldots, n,$$

we see that these conditions are transformed to the central path conditions (2.25). Therefore, the minimizers x_τ of the log barrier subproblem (2.33) are simply the x components of the central path vectors $(x_\tau, \lambda_\tau, s_\tau) \in \mathcal{C}$, so we can legitimately refer to the path of x_τ vectors as the "primal central path."

As τ decreases to zero, x_τ approaches some solution x^* of the linear program (2.1). Algorithms based on (2.33) exploit this property by finding approximations to x_τ for smaller and smaller values of τ, thereby following the primal central path to a solution of the linear program. Solution of the problem (2.33) for each value of τ can be performed, for example, by Newton's method, modified to handle the equality constraint $Ax = b$ and the condition that x remains strictly positive.

The log barrier function is usually attributed to Frisch [36]. It was studied by Fiacco and McCormick [29] in the context of nonlinear programming. (For a cogent review of its properties in this context, see Wright [145].) Algorithms based on the log barrier functions did not become popular in either the linear or the nonlinear programming community, mainly because the solution of the barrier problem becomes increasingly difficult to find as the barrier parameter τ tends to zero. Interest was revived in 1985 when the connection between Karmarkar's algorithm and log barrier algorithms was noted. Gill et al. [40] showed that Karmarkar's search direction coincides with the direction obtained by applying a Newton-like method to (2.33) for a particular choice of the barrier parameter τ.

Karmarkar's algorithm measures progress toward optimality by means of a logarithmic potential function different from the one in (2.33). In one variant of the algorithm (obtained by adapting to the standard form (2.1)), this function is

$$(n+1)\log(c^T x - Z) - \sum_{i=1}^{n} \log x_i, \tag{2.35}$$

where Z is a lower bound on the optimal objective value for (2.1), which can be updated between iterations by using information from the dual problem. Soon after Karmarkar, Renegar [112] devised an algorithm that uses Newton's method in conjunction with yet another logarithmic function, defined by

$$\begin{aligned}&\min_x \; -n\log(Z - c^T x) - \sum_{i=1}^{n} \log x_i \\ &\text{subject to } Ax = b,\ x > 0,\ c^T x < Z,\end{aligned} \tag{2.36}$$

where Z is now an *upper* bound on the optimal value Z^* of the objective $c^T x$ for (2.1). The minimizer of (2.36) is the "weighted analytic center" of the polytope defined by

$$\{x \mid Ax = b,\ x \geq 0,\ c^T x \leq Z\}, \tag{2.37}$$

where a weight of n is assigned to the constraint $c^T x \leq Z$ while unit weights are assigned to the nonnegativity constraints on x. As Z is decreased to the optimal value Z^*, the solutions of (2.36) approach a solution x^* of the linear program (2.1). In fact, each solution of (2.36) lies on the primal central path; see the exercises. Renegar's algorithm follows this path by using Newton's method to find approximate solutions of (2.36) for a sequence of decreasing values of Z. It achieves a better complexity bound than Karmarkar's method, requiring $O(\sqrt{n}\log 1/\epsilon)$ iterations to identify a primal feasible point x such that $c^T x$ is within ϵ of the optimal objective value Z^*, for given $\epsilon > 0$. No interior-point methods with better complexity than Renegar's have been identified, although numerous methods match it, as we see in Chapters 4 and 5.

In 1967, Dikin [26] proposed an algorithm that can also be motivated in terms of the log barrier function (2.33) and Newton's method. (Actually, this algorithm was rediscovered by a number of authors and only belatedly attributed to Dikin.) The algorithm finds a search direction by forming a quadratic model of the objective function in (2.33) about the current strictly feasible point x, solving the resulting quadratic program, and selecting the highest-order ($O(1/\tau)$) term from the solution. The resulting search direction Δx is

$$\Delta x = -XP_{\mathrm{Null}(AX)}(Xc), \tag{2.38}$$

where X is defined in (1.6) and $P_{\mathrm{Null}(AX)}(\cdot)$ denotes projection onto the null space of AX. Alternatively, we can view d as the step toward the point $\bar{x}$ obtained by replacing the nonnegativity constraint in (2.1) by an ellipsoidal constraint and solving

$$\min_{\bar{x}} c^T \bar{x} \text{ subject to } A\bar{x} = b,\ \|X^{-1}(\bar{x} - x)\|^2 \leq 1,$$

and then setting $\Delta x = \bar{x} - x$. This method is known as the *primal affine-scaling method* because of the affine-scaling transformation with the matrix X that is applied to the primal variables at each iteration.

Primal-Dual Methods: Development of the Fundamental Ideas

The fundamental ideas of primal-dual methods were developed between 1987 and 1991. The major causes of this burst of activity were the explicit definition of the primal-dual central path $\mathcal{C}$ and the insight into its properties provided by a number of authors. Although the transition from the log barrier viewpoint to the central path definition (2.25) is technically simple, the elevation of the dual variables λ and s from a supporting role to equal status with x gave a new perspective on pathways to the solution set.

McLinden [83] defined the central path in 1980 for a primal-dual pair of convex optimization problems and an associated nonlinear complementarity problem, proving results about existence of the path and convergence to a strictly complementary solution. Sonnevend [121] defined the primal central path as the sequence of analytic centers of the polytopes (2.37). Bayer and Lagarias [13, 14, 15] discussed both the primal and primal-dual central paths in the context of linear programming, examining these trajectories from many different mathematical points of view. Possibly the most influential paper, however, was the one written by Megiddo [85] in 1987, which many regard as the cornerstone for the field of primal-dual algorithms. Besides defining the central path $\mathcal{C}$ in the simple terms (2.25) and relating it to the log barrier function (2.33), this paper provided remarkable insight on several important issues, including higher-order approximations to the central path, the near-solution geometry of $\mathcal{C}$ (an understanding of which is essential to superlinear convergence analysis), and the generalization of the central path definition to LCPs.

Motivated by Megiddo's work, Kojima, Mizuno, and Yoshise [67] developed the first polynomial primal-dual algorithm in 1987. Their long-step path-following algorithm restricts its iterates to a central path neighborhood of the form $\mathcal{N}_{-\infty}(\cdot)$ defined in (1.16) and is closely related to Algorithm LPF of Chapter 5. Soon afterwards, the same authors [66] described a short-step path-following algorithm, in which the iterates were restricted to the tighter $\mathcal{N}_2(\cdot)$ neighborhood. Its complexity estimate of $O(\sqrt{n} \log 1/\epsilon)$ was an improvement on the $O(n \log 1/\epsilon)$ bound obtained for the long-step algorithm. A similar short-step algorithm was proposed at the same time by Monteiro and Adler [94]. In 1990, Mizuno, Todd, and Ye [92] improved on the short-step approach with their predictor-corrector algorithm, which allowed an adaptive choice of step length while retaining the same complexity bound.

A parallel stream of development took its note from the log barrier function (2.33) and Karmarkar's [57] use of a logarithmic potential function. In this line of work, logarithmic potential or barrier functions are used to generate search directions and/or monitor the progress of the algorithm. Initially, the focus was on functions of the primal variables such as (2.33), (2.35), and (2.36). Information from the dual problem was used only to update the optimal objective bound Z, if at all. For algorithms of this type, see Gonzaga [45] and Ye [154], as well as the papers of Karmarkar [57] and Renegar [112] discussed above. Introduction of the primal-dual potential function (1.18) by Tanabe [125] and Todd and Ye [131] led to a more important role for the dual variables, but some algorithms continued to update the primal and dual variables separately or by different means. Examples include algorithms described by Ye [154], Freund [34], and Gonzaga and Todd [47]. The elegant primal-dual potential reduction method of Kojima, Mizuno, and Yoshise [68] achieves a truly balanced treatment of the primal and dual variables, analogous to the primal-dual path-following algorithms of the preceding paragraph. It searches along the directions of the form (1.12), using the potential function (1.18) to choose the step length. This is the method we describe in Chapter 4.

The affine-scaling algorithm of Dikin [26] provided another focus for concentrated research in the years following 1985. This research focused mainly on the primal algorithm (with search direction (2.38)) or its dual counterpart, with different rules for choosing the step length. No polynomial complexity results are known for these algorithms, but global convergence results are proved by many authors; see, for example, Tsuchiya and Muramatsu [134], Monteiro, Tsuchiya, and Wang [96], and Tseng and Luo [133]. A primal-dual variant of the affine-scaling algorithm was proposed by Monteiro, Adler, and Resende [95], who used the pure Newton step for $F(x, \lambda, s)$ defined in (1.5a) as the search direction. As noted earlier, we can compute this direction by setting $\sigma = 0$ in (1.12). The justification for calling this direction an "affine-scaling" direction comes from the observation that its Δx component is

$$\Delta x = -DP_{\mathrm{Null}(AD)}(Dc) \tag{2.39}$$

(where $D = S^{-1/2}X^{1/2}$ as in (1.24)), which differs from the Dikin search direction (2.38) only in the choice of matrix for the affine-scaling transformation. Polynomial complexity results for primal-dual affine scaling are proven in [95], while Jansen, Roos, and Terlaky [54] define a somewhat different primal-dual affine-scaling algorithm and also prove polynomial complexity.

Many enhancements were subsequently applied to these basic algorithms, including modifications to handle infeasible initial points, ensure superlinear convergence, and improve robustness. We discuss the historical origins of these and other ideas in the relevant chapters below.

Notes and References

For a derivation of the KKT conditions from first principles, see Mangasarian [81, Chapter 2] and Schrijver [119]. Farkas's lemma is proven in both books and is used as the basis for their analysis. Fletcher [30, Chapter 9] also uses Farkas's lemma to derive first-order necessary conditions for general constrained optimization. The conditions (2.4) are a special case of these general KKT conditions.

Our proof from first principles of the existence of the central path is from Güler et al. [49]. An alternative proof can be obtained by focusing on the primal log barrier function (2.33) and showing that it has a unique optimal solution x_τ for each $\tau > 0$. As pointed out by Lewis and Overton [73, section 11], this result is in turn a consequence of the Fenchel duality theorem; see Rockafellar [116, section 31]. A proof from first principles of existence of a minimizer of the primal log barrier function is given by Wright [145] for the more general case of nonlinear convex programming.

The geometry of the central path $\mathcal{C}$ is explained in a satisfying way by Vavasis and Ye [141]. They design an algorithm that exploits the fact that $\mathcal{C}$ consists of smooth segments connected by sharp turns and show that the running time of their method depends only on the amount of storage needed for the matrix A (and not on b and c). Movies of a central path, projected into three-dimensional dual space, can be found on the World Wide Web through the home page for this book. (See the preface for the URL.) The geometry suggested by Vavasis and Ye is clearly present in these examples, with the central path taking a couple of sharp turns before finally settling down to a linear approach to the solution.

Some excellent survey papers and books are available to which the reader can refer for additional detail on the development of primal-dual interior-point methods. The 1991 monograph of Kojima et al. [65] surveyed polynomial algorithms for linear complementarity problems, which, as we note above and in Chapter 8, are closely related to primal-dual algorithms for linear programming. In a 1992 paper, Gonzaga [46] gives a comprehensive survey of developments in path-following algorithms (both primal and primal-dual) to that time. A fairly recent survey of computational issues is given by Lustig et al. [79], although additional developments have taken place since the publication of this paper and more can be expected in future. Finally, we mention a forthcoming book on interior-point methods by Ye [156], which includes extensive discussions of mathematical aspects such as geometry and complexity theory.

Exercises

1. Verify that (2.13) is the dual of (2.12) and that the solutions do indeed have the form (2.14).

2. Prove Corollary 2.7 by splitting the equality $Hd = 0$ into two inequalities and applying Lemma 2.6.

3. Suppose that $d \in \mathbb{R}^n$ is a direction for which

$$c^T d < 0, \qquad Ad = 0, \qquad d_i \geq 0 \text{ for each } i \text{ with } x_i = 0.$$

   Then for any feasible vector x for (2.1), show that $x + \alpha d$ is feasible and $c^T(x + \alpha d) < c^T x$ for all $\alpha > 0$ sufficiently small. Deduce that if such a vector d exists, the objective function $c^T x$ is unbounded below on its feasible region and hence (2.1) has no finite solution.

4. Use the results of the previous two exercises to prove that if x^* is a solution of (2.1), there exist λ and s such that conditions (2.4) are satisfied.

5. Consider the following linear program in nonstandard form:

$$\min_{w,z}\ c^T w + d^T z \qquad \text{subject to}$$
$$A_{11}w + A_{12}z \geq b_1, \qquad A_{21}w + A_{22}z = b_2, \qquad w \geq 0.$$

   (i) By splitting and introducing slack variables, convert this problem to standard form.

   (ii) Write down the dual problem for the standard-form version, and express it in a cleaner form by combining split variables and eliminating slack variables.

   (iii) By using Theorem A.1, check that the KKT conditions for the original primal problem above and your final dual problem are identical.

6. Write down standard-form linear programs in which $x \in \mathbb{R}^n$ and

   (i) $\mathcal{B} = \emptyset$;

   (ii) $\mathcal{B} = \{1, 2, \ldots, n\}$.

7. Show that the KKT conditions are necessary and sufficient for dual optimality by performing the following two steps:

   (i) By applying Theorem A.1, show that if (λ, s) is a solution of the dual problem (2.2), there is a vector x such that the KKT conditions (2.4) are satisfied.

   (ii) Given a vector (x, λ, s) that satisfies (2.4), show that any other dual feasible vector $(\bar{\lambda}, \bar{s})$ has $b^T\bar{\lambda} \leq b^T\lambda$ and therefore that the KKT conditions are also *sufficient* for dual optimality.

8. Prove the second statement in Theorem 2.3.

9. Prove that x^* defined by (2.15) has $x^*_{\mathcal{B}} > 0$.

10. We showed that the central path equations could be derived from the KKT conditions for the log barrier formulation of the primal problem (2.1). Show that the log barrier formulation of the dual problem (2.2) leads to the same system (2.25).

11. Using Theorem A.1, write down the KKT conditions for a solution of (2.36). By relating these conditions to (2.34), show that the solution of (2.36) also solves (2.33) for a certain value of τ.

12. Given a point $(x, \lambda, s) \in \mathcal{F}^o$, show that the Δx component of the primal-dual affine-scaling step obtained by setting $\sigma = 0$ in (1.12) satisfies (2.39). (Hint: Use the result that Null(B^T) is orthogonal to Range(B) for any matrix B.)

Chapter 3

Complexity Theory

In this chapter, we discuss a purely theoretical issue: basic complexity theory for linear programming. We ask such questions as, Can we find an upper bound on the time needed by an algorithm to solve a given instance of a linear programming problem? Can we estimate the time required to solve a "typical" problem instance? What properties of the problem instance does the bound depend on, and how does it depend on them?

Complexity theory is the theory of computational algorithms. It has played an important role in the history of linear programming and interior-point methods. The search for methods with better complexity than the simplex method led Khachiyan [60] to develop the ellipsoid method in 1979 and Karmarkar [57] to propose his interior-point method in 1984. Today, many of the advances in the interior-point field still are made by researchers who are experts in complexity theory rather than in the more practical aspects of mathematical programming.

Many papers in the interior-point literature state and prove results about complexity; we include analysis of this type in Chapters 4, 5, and 6. To understand these results, it helps to know some elementary complexity theory. We outline here just a few aspects of the theory, with a focus on the aspects that are relevant to primal-dual algorithms. Broader treatments of complexity theory can be found elsewhere—for example, in the book by Garey and Johnson [37]. The works of Schrijver [119] and of Vavasis [139] are also of particular interest, because they deal with the role of complexity theory in optimization.

In this chapter, we start by describing a model of computation based on rational-number arithmetic that provides a convenient framework for analyzing numerical algorithms. Complexity results for the rational-number model are sometimes referred to as *bit complexity* results. Most results in the interior-point literature are of this type; we outline the major ones later in the chapter. Next, we discuss the shortcomings of the bit complexity framework and introduce *algebraic complexity*, which is tied to a model of computation based on real arithmetic. We conclude with a general complexity theorem for path-following primal-dual methods, which is used as the basis for the results of Chapters 5 and 6.

Polynomial Versus Exponential, Worst Case Versus Average Case

Complexity theory makes a fundamental distinction between two kinds of algorithms. One kind—*polynomial algorithms*—is guaranteed to solve each problem instance in an amount of time that is polynomial in some measure of the problem size. In complexity theory, the term *efficient* often is used to denote polynomial algorithms. The other kind—*exponential algorithms*—may require an amount of time that is exponential in the problem size. Exponential bounds grow at a spectacularly faster rate than polynomial bounds. For example, the number of elementary particles in the visible universe is estimated at between e^{78} and e^{80}, yet a problem of size 80 would be regarded as small by almost any measure.

For linear programming, the simplex algorithm is an exponential algorithm, but an especially interesting one because it requires much less computing time in practice than might be expected from the theory. Khachiyan's ellipsoid method [60] is polynomial but, paradoxically, much slower than the simplex method in practice. Karmarkar's method [57] and most other interior-point methods are polynomial, and the practical performance of some of these algorithms is also very good.

Through much of its history, complexity theory has been concerned with the *worst-case* behavior of algorithms, that is, with finding an upper bound on the time required to solve any problem instance from the chosen class. Because some algorithms show a discrepancy between the worst-case bound and performance on typical problem instances, a fair amount of recent research has focused on *average-case* performance. The general technique of average-case analysis is to impose a certain probability distribution on the class of problem instances and then to identify the "expected" performance of the algorithm under this distribution. As we observed in the preceding paragraph, the simplex method exhibits a wide discrepancy between worst-case and average-case performance: the worst-case behavior is exponential in the dimension n, while the average case is just polynomial in n. A gap also exists for interior-point methods: the worst case is polynomial in n, while the average case depends only on a fractional power of n or even just on $\log n$. For the ellipsoid method, on the other hand, the worst case and the average case are much the same.

For simplicity, we restrict our discussion in this chapter to the worst-case behavior. We refer the interested reader to Borgwardt [18] for an average-case analysis of the simplex method and to Anstreicher et al. [11] for an average-case analysis of interior-point methods for linear programming.

Storing the Problem Data: Dimension and Size

The *problem size* is the amount of computer storage required to store all the data that define the problem instance. In the case of linear programming, this *problem data* consists of three objects: the matrix A and the vectors b and c. If all the entries in (A, b, c) are integers, the data can be stored exactly in a computer, since an integer x can be represented by using $\lceil \log_2 |x| \rceil + 1$ bits of storage. If the problem data are rational, they can still be stored exactly, since each rational number can be described by two integers: its numerator and denominator. Irrational numbers, however, cannot be stored exactly with standard binary representations. This is the reason that bit complexity results apply only to problems with rational data.

If the data (A, b, c) are rational, we can easily convert them to integer by identifying the least common denominator and multiplying through. Hence, we assume for the analysis below that the problem data (A, b, c) are integer.

Since the matrix A is $m \times n$ and the vectors b and c have m and n entries, respectively, the total number of integers in the data string is $mn + m + n$. Additional separator symbols are required to indicate where each integer starts and finishes. (For simplicity, we assume that each of these symbols is represented by a single "special" bit.) If L_0 bits are required to store the integers themselves, the total length of the data string—the problem size—is

$$L \stackrel{\text{def}}{=} L_0 + (mn + m + n). \tag{3.1}$$

The use of the quantity "L" is standard (indeed, ubiquitous) in the interior-point literature.

Apart from L, the quantity that appears most often in bit complexity analysis is n, the number of columns in A (2.1). In linear programming, we use n to measure the *problem dimension*. Some complexity results also refer to m, the number of rows in A. If we assume that A has full row rank, however, we have $m \leq n$, so the various bounds still hold (and are simpler) if we replace m by n everywhere.

The Turing Machine and Rational Arithmetic

The Turing machine is a conceptual device that is used in complexity theory as a model of the computing process. A brief description is given in Appendix A. The power of the Turing machine lies in its simplicity and generality, but for the complexity analysis of numerical algorithms it is easier to work with higher-level models of computation. Complexity results in mathematical programming traditionally are stated with reference to the *rational-number model.* Rational arithmetic can be carried out by performing integer arithmetic on the numerators and denominators of the operands. For instance, the operation

$$\frac{151}{210} \div \frac{79}{23} = \frac{3473}{16590} \tag{3.2}$$

is carried out via the two integer multiplications $151 \times 23 = 3473$ and $210 \times 79 = 16590$. It is not too hard to imagine how integer arithmetic like this could be performed on a Turing machine. The machine could simply be programmed to go through the same pencil-and-paper procedures for integer multiplication, addition, and subtraction that most of us remember from primary school (and can still perform in an emergency). In fact, the entire rational-number model of computation can be emulated on a Turing machine—not just the arithmetic itself but also the process of retrieving the operands from storage and storing the result. The number of steps needed by the Turing machine to perform each unit of rational computation is a polynomial function of the number of bits required to store the operands, so the emulation is efficient in the sense of complexity theory.

Note that the result of the rational arithmetic operation may require considerable extra storage space, possibly as much as the total space required to store the operands; witness the example (3.2).

Primal-Dual Methods and Rational Arithmetic

It is obvious that the simple arithmetic operations of $+$, $-$, $\times$, and $\div$ performed on two rational arguments produce a rational result, whereas the square root operation does not preserve rationality. The process of moving from one primal-dual iterate to the next involves large-scale computations such as matrix multiplications and the solution of linear systems. Since almost all these computations are composed from the basic $+$, $-$, $\times$, and $\div$

scalar operations, they preserve rationality. For instance, we can show by an application of Cramer's rule (see below) that if the linear system (1.12) (or (1.20)) has a rational coefficient matrix and rational right-hand side, the solution is also rational.

One operation in the algorithms of Chapters 4, 5, and 6 generally does not preserve rationality: determination of the step length α. Often, to find α, we must solve quadratic or quartic equations, a process that involves taking square roots. In all cases, however, the algorithms can be modified to allow for an approximate value of α to be used instead—a rational value that is found by using only rational arithmetic yet is accurate enough to produce the same theoretical behavior of the algorithm as the exact α. We conclude, then, that if the linear program has integer (or rational) data (A, b, c) and the primal-dual method is started from a rational point (x^0, λ^0, s^0), all iterates (x^k, λ^k, s^k) are rational.

One possible concern is the size of the iterates (x^k, λ^k, s^k). Could this vector require more and more storage as k increases—much more, perhaps, than the original length L of the input string? The concern is legitimate because, as we noted earlier, the space required to store the result of a rational operation may be equal to the total space required by its two operands, giving the potential for storage to increase exponentially with operation count. A standard result (see, for example, Schrijver [119, Corollary 3.2b]) is that the size of a solution to a linear system of equations is bounded by a polynomial in the sizes of its matrix and right-hand side. Hence, the size of $(\Delta x^k, \Delta\lambda^k, \Delta s^k)$, calculated from (1.12), is bounded by a polynomial in the sizes of the original data (A, b, c) and the current iterate (x^k, λ^k, s^k). To prevent the size of (x^k, λ^k, s^k) increasing exponentially with k, however, algorithms must allow some rounding of the iterates to $O(L)$ bits. That is, they must allow for inexact arithmetic in the solution of the system (1.12) and in other operations. Khachiyan [60], Karmarkar [57], and Renegar [112] all devote considerable effort to showing that their respective algorithms allow inexact arithmetic to be performed without compromising their convergence and complexity properties.

The number of rational arithmetic operations performed at each primal-dual step depends polynomially on the dimension n; typically, it is proportional to n^3. Dependence of operation count on the size of the data L is marginally worse than linear but certainly still polynomial. (Karmarkar [57, p. 377] mentions a factor of $L(\log L)(\log\log L)$ for factorization of the coefficient matrix in his algorithm.)

At this point, we have determined that each iteration of an interior-point algorithm requires a polynomial number of rational operations and that each rational operation can be implemented in polynomial time on the Turing machine. Moreover, the storage requirements at each iteration are not appreciably greater than the size of the problem data. Hence, *if we can show that the interior-point algorithm finds a solution in a polynomial number of iterations, we can conclude that the whole algorithm will run in polynomial time on the Turing machine.* Finding the bound on the number of iterations is the focus of the complexity analysis in most interior-point papers.

Linear Programming and Rational Numbers

We assume in this section and the next that the integer matrix A has full row rank. If not, we can perform QR factorization as in Chapter 2 to obtain an equivalent problem for which the full rank assumption holds. Since the QR factorization requires $O(n^3)$ operations, it does not compromise polynomiality of the overall solution process.

Recall the definition of *basic feasible points* from Chapter 2: They are points x that satisfy the primal feasibility conditions $Ax = b$ and $x \geq 0$ and can be expressed in the form

$$x = \Pi \begin{bmatrix} x_B \\ 0 \end{bmatrix} = \Pi \begin{bmatrix} B^{-1}b \\ 0 \end{bmatrix}, \tag{3.3}$$

where Π is an $n \times n$ permutation matrix and B is a $m \times m$ nonsingular column submatrix of A.

As we mentioned in Chapter 2, each basic feasible point is a vertex of the primal feasible polygon $\{x \mid Ax = b,\ x \geq 0\}$, and vice versa. When the data (A, b, c) are integer, with length L defined by (3.1), we can say more about the properties of these vertices.

Lemma 3.1 *When the problem data (A, b, c) are integer with length L, the vertices of the primal and dual feasible polygons defined by*

$$\{x \mid Ax = b,\ x \geq 0\}, \qquad \{(\lambda, s) \mid A^T\lambda + s = c,\ s \geq 0\}$$

are rational. Moreover, the nonzero components of x and s for these vertices are bounded below by 2^{-L}.

Proof. The result follows from a lemma of Vavasis [139, Lemma 3.1]. Since x_B in (3.3) is the solution of $Bx_B = b$, where (B, b) can be written in at most L_0 bits, the cited result implies that the components of x_B

are rational, with a common denominator Δ whose magnitude is at most $2^{L_0+m\log m}$. Since each entry of x_B is an integer multiple of $1/\Delta$, its nonzero entries must be bounded below by $1/|\Delta|$, which in turn is bounded below by $2^{-(L_0+m\log m)}$. Since $m \le n$, we have

$$L_0 + m\log m \le L_0 + (mn + m + n) = L,$$

so the result is proved for vertices of the primal polygon.

Vertices of the dual feasible polygon have the form

$$\lambda = B^{-T}c_B, \qquad s = \Pi \begin{bmatrix} 0 \\ c_N - N^T\lambda \end{bmatrix} \ge 0, \tag{3.4}$$

where N contains the columns of A that do not belong to B in (3.3), c is partitioned accordingly into c_B and c_N, and Π is a permutation matrix as before. Applying the cited lemma again, we find that λ satisfying (3.4) is a rational vector whose entries have a common denominator Δ, where $|\Delta| \le 2^L$. Since the entries of c_N and N are integers, the entries of s in (3.4) are integer multiples of $1/\Delta$. The argument that we used for x above can be applied to s to show that the nonzero components of s are bounded below by 2^{-L}. □

Moving to a Solution from an Interior Point

Interior-point algorithms typically generate a sequence of strictly feasible points $\{(x^k, \lambda^k, s^k)\}$ for which x^k and s^k are both strictly positive. None of these iterates can be exact solutions of the linear program because we have $x_i^k s_i^k > 0$ for each $i = 1, 2, \ldots, n$, violating the complementarity condition (1.4c). The iterates converge to the solution set without ever quite reaching it, and for many practical purposes an advanced iterate ((x^k, λ^k, s^k) for k sufficiently large) serves quite well as an approximate solution. For other practical applications, however, and for the theoretical purposes of this chapter, an exact solution $(x^*, \lambda^*, s^*) \in \Omega$ is required.

To find an exact solution, we need a *finite termination* procedure that takes us directly from a sufficiently advanced interior-point iterate to a point in Ω. We also need to know just *how* advanced the iterate needs to be, so we can be sure that the required number of interior-point iterations is not too great.

Procedure FT, which we outline in Chapter 7, accomplishes this task by projecting the current point (x^k, λ^k, s^k) onto the nearest point (x^*, λ^*, s^*)

that satisfies the first three KKT conditions (1.4a), (1.4b), (1.4c). We show in Chapter 7 that, after the duality measure μ_k falls below some threshold $\bar{\mu} > 0$, the projected point (x^*, λ^*, s^*) will also satisfy the nonnegativity condition $(x^*, s^*) \geq 0$ (the fourth KKT condition) and will therefore belong to Ω. The threshold $\bar{\mu}$ is proportional to the square of the quantity $\epsilon(A, b, c)$ defined by

$$\epsilon(A, b, c) = \min \left(\min_{i \in \mathcal{B}} \sup_{x^* \in \Omega_P} x_i^*, \min_{i \in \mathcal{N}} \sup_{(\lambda^*, s^*) \in \Omega_D} s_i^* \right), \tag{3.5}$$

where $\mathcal{B} \cup \mathcal{N}$ is the partition of the index set $\{1, 2, \ldots, n\}$ described in Chapter 2. It follows from Lemma 3.1 that if the problem data (A, b, c) are integer with size L, we must have

$$\epsilon(A, b, c) \geq 2^{-L}. \tag{3.6}$$

To verify this result, we must show first that

$$\sup_{x^* \in \Omega_P} x_i^* \geq 2^{-L} \qquad \text{for all } i \in \mathcal{B}. \tag{3.7}$$

If not, we have $0 \leq x_i^* < 2^{-L}$ for all *vertex* solutions $x^* \in \mathcal{P}$. Since the nonzero components of a vertex/basic feasible point are no smaller than 2^{-L}, it follows that $x_i^* = 0$ for all vertex solutions. Since the vertex solutions are the extreme points of Ω_P, every primal solution is a convex combination of vertex solutions, so we must have $x_i^* = 0$ for all $x^* \in \Omega_P$. This contradicts the definition of $\mathcal{B}$ (2.11a), proving (3.7). In a similar way, we can show that $\sup_{(\lambda^*, s^*) \in \Omega_D} s_i^* \geq 2^{-L}$ for all $i \in \mathcal{N}$, so (3.6) is proved.

The other factors in the threshold $\bar{\mu}$ have a similar dependence on L, so we have that

$$\bar{\mu} \geq 2^{-\bar{t}L} = e^{-tL}, \tag{3.8}$$

where $\bar{t}$ is a moderate multiplier (no more than 10) and $t = \bar{t} \log_e 2$.

To summarize our (theoretical) approach for finding an exact solution: We use an interior-point algorithm to generate a sequence $\{(x^k, \lambda^k, s^k)\}$ with $\mu_k \downarrow 0$. We terminate this algorithm at the first iterate k that satisfies the condition

$$\mu_k \leq e^{-tL}, \tag{3.9}$$

since this value guarantees that $\mu_k \leq \bar{\mu}$. Finally, we activate Procedure FT to find an exact solution.

In practice, the computer never accepts input in the form described above and never performs rational arithmetic. Moreover, no practical algorithm can iterate until the very stringent criterion (3.9) is satisfied, since the errors associated with floating-point computations interfere long before this point is reached. Still, a practical implementation of the two-phase procedure would be similar to the conceptual version outlined in the preceding paragraph. The principal difference would be that we would activate Procedure FT at much larger values of μ_k than the threshold condition (3.9) would suggest, and in most cases, we would terminate at an exact solution (modulo roundoff error) in many fewer iterations than this idealized, worst-case analysis would indicate.

Other applications demand even more than an exact solution: they require a *vertex* solution, one for which x^* is a vertex of the primal feasible polygon and (λ^*, s^*) is a vertex of the dual feasible polygon. Such solutions are useful when the linear program is a subproblem in an integer programming algorithm. The heuristics for pruning the tree in a branch-and-bound algorithm, for instance, operate better when the linear programming solver returns a vertex solution. Fortunately, this requirement can be met without compromising polynomiality. A procedure due to Megiddo [86], which we describe in Chapter 7, obtains a vertex solution from any primal-dual solution in polynomial time.

Complexity of Simplex, Ellipsoid, and Interior-Point Methods

One of the main purposes of complexity theory is to differentiate between efficient and inefficient algorithms. In the field of linear programming, however, the correlation between good complexity properties and good practical performance is not particularly strong.

The simplex method may require a number of iterations that is exponential in the dimension n, so it is certainly not a polynomial algorithm. The Klee–Minty example [61], for instance, requires 2^n simplex iterations. Such behavior is not at all typical, however; almost all problem instances require fewer than $3n$ iterations. The average-case complexity analysis of Borgwardt [18] has shed some light on this discrepancy between practical experience and worst-case complexity analysis.

Khachiyan's [60] ellipsoid algorithm was the first polynomial algorithm for linear programming, requiring at most $O(n^2L)$ iterations to solve the problem. Unfortunately, this worst-case bound actually is achieved on most problems, and the algorithm is much slower than the simplex method on all but the smallest problems.

Karmarkar's polynomial algorithm [57] has an iteration bound of $O(nL)$. This complexity was subsequently improved to $O(\sqrt{n}L)$ by Renegar [112], still the best known bound today. Besides their good complexity properties, these algorithms perform much better than Khachiyan's algorithm and are not outrageously slower than the simplex method. Both good complexity properties and good practical performance are embodied in primal-dual methods, the important subclass of interior-point methods that we focus on in this book.

Within the class of interior-point methods, the exponent of n in the worst-case bound does not have much bearing on practical behavior. Most algorithms that achieve the best known complexity of $O(\sqrt{n}L)$ are slow, while other algorithms for which the iteration bound is $O(n^2L)$ are considerably faster. Many practical implementations of primal-dual algorithms include heuristics that make them ineligible for any kind of complexity analysis, although they are all closely related to polynomial algorithms.

Polynomial and Strongly Polynomial Algorithms

Recall that polynomial algorithms have a running time that depends polynomially on the size of the problem, which we measure in the case of integer data by the length L of the input string (3.1). Since $L > n$, the product $n^\omega L$ is polynomial in L for any $\omega \geq 0$, so any interior-point algorithm that converges in $O(n^\omega L)$ iterations is a polynomial algorithm.

An algorithm is said to be *strongly polynomial* if it is polynomial and if the total number of arithmetic operations it performs is bounded by a polynomial in the dimension of the problem—in our case, n. Interior-point methods are not strongly polynomial for problems with integer data, since the number of arithmetic operations depends on L, which may be arbitrarily larger than n.

However, certain algorithms *are* strongly polynomial for important subclasses of linear programming. Tardos [127] described an algorithm for which the running time is polynomial in the size of A alone (that is, there is no dependence on b and c). For network flow problems, A contains only zeros and ± 1 elements, so it can be stored in $O(n^2)$ bits. For these problems, then, Tardos's method is strongly polynomial. Furthermore, Vavasis and Ye [141] described an interior-point algorithm whose running time depends solely on a certain condition measure of the matrix A. By showing that this condition measure is polynomial in the size of A, they recovered an algorithm that is strongly polynomial on the same class of problems as Tardos's method.

Beyond the Turing Machine Model

As a model for numerical computation, the Turing machine/rational-number model has deficiencies that are particularly apparent in its application to linear programming. Recent work has focused on an alternative model of computation based on real numbers due to Blum, Shub, and Smale [16]. This *real-number model* differs from the Turing machine model in that its "alphabet" is the set of real numbers. A problem instance with real data (A, b, c) can be stored on the input tape by assigning one real number per cell, so that the length of the input string is $mn + m + n$. That is, the problem size L in the real-number model is $L = mn + m + n$. This model assumes that the algorithms perform *exact* real arithmetic.

Because it gives a more realistic picture of how data are input, stored, and operated on, the real-number model is closer to computational practice than the rational-number model. The real-number model overcomes one outstanding deficiency of the rational-number model: the dependence of the complexity results on the size L of the (integer or rational) problem data (A, b, c). Bit complexity analysis tells us that when L is large, the rational arithmetic takes longer to perform and the convergence tolerance (3.9) is tighter, so it suggests that problems with larger L take longer to solve. It is, however, simple to construct two rational linear programs for which the numerical values in (A, b, c) are extremely close but the space required to store the data differs enormously. In the rational-number model, the times required to solve these two problem instances would be very different. In practice, however, we would expect our computer code to spend similar amounts of time solving the two problems, because of their close numerical similarity. In the real-number model, the complexity bounds for these two problems would be the same, because their dimensions are identical.

The real-number model is simple and universal enough to be used as the basis of an extensive theory. However, it falls short of computational practice in two ways. First, on actual computers, we can store only a subset of the reals known as *floating-point numbers*. The actual data (A, b, c) usually must be truncated or rounded to the nearest floating-point number before it can be stored and operated on. The rounding process introduces perturbations that can significantly affect the solution. The second difference with computing practice is that arithmetic operations on actual computers are not exact. Computers perform floating-point arithmetic, in which roundoff error is introduced at almost every operation. A model for floating-point computation and for the error analysis of algorithms was proposed by Wilkinson in the 1960s [143, 144]. Wilkinson's model has proved to be an extremely useful tool in the numerical analysis of algorithms for linear algebra (matrix factorizations, triangular substitutions, and so on), which are not excessively complicated procedures. Because of its low-level nature, however, Wilkinson's model is more difficult to apply in more complex situations such as optimization algorithms. Some recent research has incorporated the effects of roundoff and inexactness into the convergence/complexity theory of interior-point methods (see, for example, Vera [142]), but our understanding of these issues is far from complete.

Putting aside L, can we identify other properties of the data that strongly influence the performance of the algorithm? A related question is, what properties of the data make the problem stable under perturbations due to roundoff errors and floating-point representations? The search for these properties is a current focus of research; for an introduction, see Renegar [114, 115]. As in the rational-number model, the quantity $\epsilon(A, b, c)$ defined in (3.5) appears to be an important indicator of the "difficulty" of the problem instance in the real-number model.

More on the Real-Number Model and Algebraic Complexity

Complexity results based on the real-number model are sometimes called *algebraic complexity* results, to distinguish them from the bit complexity results that are associated with the rational-number model.

Interior-point methods are not polynomial in the sense of algebraic complexity. Recall that in the rational-number model, a finite termination procedure is used to move to an exact solution once the duality measure μ drops below the threshold (3.9). The value e^{-tL} of this threshold may be tiny, but it is at least nonzero and dependent only on the length of the data string and not on the numerical values. Since the quantity $\epsilon(A, b, c)$ may be arbitrarily small in the real-number model, we may have to reduce μ to an arbitrarily small value before the finite termination procedure becomes effective. Hence, the interior-point algorithm may require arbitrarily many iterations. Existence of a polynomial algorithm for linear programming, in the sense of algebraic complexity, remains an open question.

Unlike bit complexity results, algebraic complexity results do not look for bounds on the number of iterations required to identify an exact solution. Rather, they identify an upper bound on the number of iterations needed to reduce μ_k below some small prescribed threshold ϵ. This criterion often is used in actual implementations; an iterate (x^k, λ^k, s^k) that satisfies $\mu_k \leq \epsilon$

and is feasible serves perfectly well as an approximate solution of the linear program for most purposes.

Despite these fundamental differences, little effort is required to change bit complexity results for interior-point methods into algebraic complexity results. Generally, the only change we need to make is to replace L by $\log 1/\epsilon$ in the analysis.

In the next section we state and prove a general complexity result for a primal-dual path-following method. This result is typical of many in the literature, and we refer to it repeatedly in later chapters.

A General Complexity Theorem for Path-Following Methods

Theorem 3.2 shows that if the reduction in μ at each iteration depends on the dimension n in a certain way, and if the initial duality measure is not too large, the algorithm has polynomial complexity.

Theorem 3.2 *Let $\epsilon \in (0,1)$ be given. Suppose that our algorithm for solving (2.4) generates a sequence of iterates that satisfies*

$$\mu_{k+1} \leq \left(1 - \frac{\delta}{n^\omega}\right)\mu_k, \qquad k = 0,1,2,\ldots, \tag{3.10}$$

for some positive constants δ and ω. Suppose too that the starting point (x^0, λ^0, s^0) satisfies

$$\mu_0 \leq 1/\epsilon^\kappa \tag{3.11}$$

for some positive constant κ. Then there exists an index K with

$$K = O\left(n^\omega |\log \epsilon|\right)$$

such that

$$\mu_k \leq \epsilon \qquad \textit{for all } k \geq K.$$

Proof. By taking logarithms of both sides in (3.10), we obtain

$$\log \mu_{k+1} \leq \log\left(1 - \frac{\delta}{n^\omega}\right) + \log \mu_k.$$

By repeatedly applying this formula and using (3.11), we have

$$\log \mu_k \leq k \log\left(1 - \frac{\delta}{n^\omega}\right) + \log \mu_0 \leq k \log\left(1 - \frac{\delta}{n^\omega}\right) + \kappa \log \frac{1}{\epsilon}.$$

The estimate for the log function

$$\log(1+\beta) \le \beta \qquad \text{for all } \beta > -1$$

(see Lemma 4.1) implies that

$$\log \mu_k \le k\left(-\frac{\delta}{n^\omega}\right) + \kappa \log \frac{1}{\epsilon}.$$

Therefore, the convergence criterion $\mu_k \le \epsilon$ is satisfied if we have

$$k\left(-\frac{\delta}{n^\omega}\right) + \kappa \log \frac{1}{\epsilon} \le \log \epsilon.$$

This inequality holds for all k that satisfy

$$k \ge K = (1+\kappa)\frac{n^\omega}{\delta} \log \frac{1}{\epsilon},$$

so the proof is complete. □

Most of the effort in the analysis of algorithms goes into showing that the crucial condition (3.10) is satisfied, as we see in subsequent chapters.

Notes and References

Additional discussion of the Turing machine can be found in Garey and Johnson [37, p. 23] and Vavasis [139, pp. 17–19].

Karmarkar gave few details of the complexity analysis in his seminal paper [57], but he did include a finite termination procedure and a reformulation/initialization procedure to guarantee polynomial complexity.

Recent work on incorporating the effects of floating-point computations and roundoff error into complexity theory can be found in Vera [142]. Renegar [113] extends the complexity analysis of a primal potential-reduction algorithm to allow inexact computation of the step vector at each iteration. Wright's analysis [149, 148] has a more numerical flavor. He shows that the step equations can be solved to adequate accuracy by standard factorization algorithms and that the asymptotic convergence properties of interior-point algorithms are not affected too severely by roundoff error. Similar results along these lines are proved by Forsgren, Gill, and Shinnerl [31].

Exercises

1. Suppose that the constraints $Ax = b$ in (2.1) are consistent and that A does not have full rank. Show by using the QR factorization (see Appendix A) that we can replace $Ax = b$ with an equivalent set of constraints in which the coefficient matrix *does* have full row rank.

2. A linear program with integer data has solutions that contain only rational components. However, if the solution set Ω_P contains more than a single point, then it contains irrational vectors. Why?

3. Why do the definitions of polynomial and strongly polynomial coincide in the Blum–Shub–Smale real-number model?

4. Show that the result of Theorem 3.2 is not affected if we replace (3.10) with

$$\mu_{k+2} = \mu_{k+1} \le \left(1 - \frac{\delta}{n^\omega}\right) \mu_k, \qquad k = 0, 2, 4, \ldots . \tag{3.12}$$

   (This modification is used in the proof of Theorem 5.9.)

Chapter 4

Potential-Reduction Methods

Potential functions have always played a central role in the development of interior-point algorithms. Karmarkar's algorithm [57], when adapted for the standard-form linear program (1.1), makes use of a logarithmic potential function of the form

$$\rho \log(c^T x - Z) - \sum_{i=1}^{n} \log x_i, \tag{4.1}$$

where $\rho = n + 1$ and Z is a lower bound on the optimal objective value. Karmarkar proved convergence and complexity results by showing that this function is decreased by at least a constant at each step. In subsequent work, researchers devised algorithms based on variants of this potential function, and also the log barrier function (2.33) and the logarithmic function (2.36) used by Renegar [112], none of which depend explicitly on the dual variables. The complexity analysis of these algorithms depends crucially on the clever devices used to update their parameters (the barrier parameter μ in (2.33) and the bound Z in (4.1) and (2.36)). See, for example, the Newton-log-barrier algorithm discussed by Gill et al. [40] and Renegar's algorithm [112] discussed in Chapter 2. In some cases, the parameter update makes use of information from the dual problem, as in the algorithms of Todd and Burrell [129] and Gonzaga [45], but most of the action takes place in the space of primal variables.

The turn toward primal-dual path-following algorithms in 1987 caused researchers to consider potential functions and potential reduction algorithms in which the primal and dual variables played a more balanced role. The primal-dual potential function of Tanabe [125] and Todd and Ye [131], defined by

$$\Phi_\rho(x,s) = \rho \log x^T s - \sum_{i=1}^{n} \log x_i s_i \tag{4.2}$$

for some parameter $\rho > n$, was crucial to the development of potential reduction algorithms after 1988. The relationship between this function and the primal potential function (4.1) is not difficult to see. From duality theory, we know $b^T\lambda$ to be a lower bound on the objective for any dual feasible point (λ, s) (see (2.7)). Hence we can set $Z = b^T\lambda$ in (4.1) and use the relationship $x^T s = c^T x - b^T\lambda$ (2.5) to replace $c^T x - Z$ with $x^T s$. In the second term, we have simply added the summation $\sum \log s_i$, setting up a barrier term for the constraint $s \geq 0$ to match the existing barrier term for the primal constraint $x \geq 0$. The domain of the function (4.2) is the set of all vector pairs (x, s) such that

$$(x, \lambda, s) \in \mathcal{F}^o \qquad \text{for some } \lambda \in \mathbb{R}^m. \tag{4.3}$$

Some algorithms based on Φ_ρ treat the primal and dual variables in distinct ways, using the potential function only to monitor progress and prove convergence. Some, such as algorithms described by Ye [154] and Freund [34], perform line searches in the primal variables and periodically update the dual variables by other means. Others, such as the algorithm of Gonzaga and Todd [47], take steps in either the primal and dual space at each iteration, using certain measures of centrality to choose between the two alternatives. In this chapter, we describe an elegant and simple algorithm due to Kojima, Mizuno, and Yoshise [68], which takes primal-dual steps of the form (1.12) and uses Φ_ρ to choose the step length. We show that this algorithm matches the best known complexity bound for any interior-point algorithms, requiring $O(\sqrt{n}|\log \epsilon|)$ iterations to reduce the duality gap $x^T s$ below a given threshold $\epsilon > 0$.

By rewriting (4.2) as

$$\Phi_\rho(x,s) = (\rho - n)\log x^T s + \Phi_n(x,s) \tag{4.4a}$$

$$= (\rho - n)\log x^T s - \sum_{i=1}^{n} \log \frac{x_i s_i}{x^T s/n} + n \log n, \tag{4.4b}$$

we see that the function Φ_ρ balances two considerations: duality (which is better if $x^T s$ is small) and centrality (better if no $x_i s_i$ is much smaller than the average value $\mu = x^T s/n$). The function Φ_ρ acts as a barrier function as $(x, \lambda, s) \in \mathcal{F}^o$ moves toward any point for which $x_i s_i = 0$ but $x^T s > 0$. In this case, the first term in (4.4b) stays bounded, whereas the second term blows up, causing Φ_ρ to approach $+\infty$. On the other hand, Lemma 4.2 shows that the second term $\Phi_n(x,s)$ in (4.4a) has a lower bound. Therefore, Φ_ρ approaches $-\infty$ only if its first term approaches $-\infty$; that is, $\mu \downarrow 0$. This observation motivates the potential-reduction algorithm, which generates a sequence $(x^k, \lambda^k, s^k) \in \mathcal{F}^o$ such that $\Phi_\rho(x^k, s^k) \downarrow -\infty$, thereby driving the duality measure μ_k to zero and forcing the sequence to optimality.

In the remainder of this chapter, we start by specifying Algorithm PR, which is the primal-dual potential-reduction algorithm of Kojima, Mizuno, and Yoshise [68]. We then take a close look at the properties of Φ_ρ and its relationship to the duality measure μ, a relationship that holds the key to the global convergence and polynomial complexity results. Next, we show that the particular parameters σ_k and α_k used by Algorithm PR force a significant reduction in Φ_ρ at each iteration, thereby forcing the sequence of iterates to optimality. Finally, we comment on the centrality of the iterates and on "practical" choices of σ_k and α_k.

A Primal-Dual Potential-Reduction Algorithm

Algorithm PR is a special case of Framework PD from Chapter 1. It chooses a constant value $\sigma_k = n/\rho$ for the centering parameter, where $\rho > n$ is the parameter from (4.2). The step length α_k is chosen to minimize $\Phi_\rho(\cdot)$ along the computed search direction while keeping the $k+1$st iterate strictly feasible. Because of this feasibility requirement, the upper bound on step length from the point $(x, \lambda, s) \in \mathcal{F}^o$ along the search direction $(\Delta x, \Delta\lambda, \Delta s)$ is given by

$$\alpha_{\max} = \sup\{\alpha \in [0,1] \,|\, (x,s) + \alpha(\Delta x, \Delta s) \geq 0\}. \tag{4.5}$$

We are now ready to specify the algorithm.

Algorithm PR
Given $\rho > n$ and $(x^0, \lambda^0, s^0) \in \mathcal{F}^o$;
for $k = 0, 1, 2, \ldots$
set $\sigma_k = n/\rho$ and solve (1.13) to obtain $(\Delta x^k, \Delta\lambda^k, \Delta s^k)$;
calculate $\alpha_{\max}$ from (4.5), and choose the step length α_k from

$$\alpha_k = \arg \min_{\alpha \in (0, \alpha_{\max})} \Phi_\rho(x^k + \alpha\Delta x^k, s^k + \alpha\Delta s^k); \tag{4.6}$$

set $(x^{k+1}, \lambda^{k+1}, s^{k+1}) = (x^k, \lambda^k, s^k) + \alpha_k(\Delta x^k, \Delta\lambda^k, \Delta s^k)$;
end (for).

The analysis in this chapter shows that the potential function Φ_ρ is reduced by at least a constant amount (0.15, to be specific) at every iteration. The particular choices of σ_k and α_k in Algorithm PR are a little too conservative to be practical. However, we are free to make more aggressive choices (σ_k closer to zero, α_k closer to $\alpha_{\max}$), provided that we meet the essential requirement of reducing Φ_ρ by at least a fixed amount at every step. In this sense, the algorithm and its analysis are relevant to computational practice. We return to this point at the end of the chapter.

Reducing Φ_ρ Forces Convergence

We start by showing that $\Phi_\rho(x^k, s^k) \downarrow -\infty$ implies $\mu_k \downarrow 0$. This claim is proven in the three lemmas below—Lemmas 4.1, 4.2, and 4.3—which are short and uncomplicated. The final result—Theorem 4.4—gives the fundamental complexity result for algorithms based on the potential function Φ_ρ. It is the potential-reduction analogue of Theorem 3.2, the basic complexity result for path-following methods.

Our first result is a purely technical lemma about approximations to the natural log function. Part (i) is left as an exercise, whereas part (ii) is proven in Appendix A.

Lemma 4.1

(i) *We have* $\log(1+\beta) \le \beta$ *for all* $\beta > -1$, *with equality if and only if* $\beta = 0$.

(ii) *For any* $t \in \mathbb{R}^n$ *with* $\|t\|_\infty \le \tau < 1$, *we have*

$$-\sum_{i=1}^{n} \log(1+t_i) \le -e^T t + \frac{\|t\|^2}{2(1-\tau)}.$$

The next result proves our earlier claim that Φ_n, the function used in (4.4a), is bounded below.

Lemma 4.2 *For a positive vector pair* $(x, s) > 0$, *we have*

$$\Phi_n(x, s) \ge n \log n,$$

with equality if and only if $XSe = (x^T s/n)e = \mu e$.

Proof. From (4.4) and Lemma 4.1(i), we have

$$\Phi_n(x,s) - n\log n = -\sum_{i=1}^{n} \log \frac{x_i s_i}{x^T s/n} \geq -\sum_{i=1}^{n} \left(\frac{x_i s_i}{\mu} - 1 \right) = -(n-n) = 0.$$

From Lemma 4.1(i) again, we have equality if and only if

$$\frac{x_i s_i}{\mu} = 1, \qquad i = 1, 2, \ldots, n,$$

so the proof is complete. □

The next result proves that Φ_ρ is not bounded below on its domain and derives the key relationship between Φ_ρ and μ.

Lemma 4.3

(i) *Φ_ρ is unbounded below on its domain.*

(ii) *For any $(x, \lambda, s) \in \mathcal{F}^o$, we have*

$$\mu \leq \exp\left(\Phi_\rho(x,s)/(\rho - n)\right), \tag{4.7}$$

where $\mu = x^T s/n$.

Proof. From Theorem 2.8, the result about existence of the central path $\mathcal{C}$, we know that for any $\mu > 0$ there is a strictly feasible point $(x_\mu, \lambda_\mu, s_\mu)$ such that $(x_\mu)_i(s_\mu)_i = \mu$ for all $i = 1, 2, \ldots, n$. Substituting this point in (4.4a), we obtain

$$\begin{aligned} \Phi_\rho(x_\mu, s_\mu) &= (\rho - n)\log x_\mu^T s_\mu + \Phi_n(x_\mu, s_\mu) \\ &= (\rho - n)\log(n\mu) + n\log n, \end{aligned}$$

which approaches $-\infty$ as $\mu \downarrow 0$.

For part (ii), we use (4.4) and Lemma 4.2 to write

$$\begin{aligned} \Phi_\rho(x,s) &= (\rho - n)\log x^T s + \Phi_n(x,s) \\ &\geq (\rho - n)\log \mu + (\rho - n)\log n + n\log n \\ &\geq (\rho - n)\log \mu, \end{aligned}$$

from which the result follows immediately. □

It follows from Lemma 4.3(ii) that if we can generate a sequence of iterates $(x^k, \lambda^k, s^k) \in \mathcal{F}^o$ for which $\Phi_\rho(x^k, s^k) \downarrow -\infty$, then $\mu_k \to 0$. In fact, because of (2.6) and (2.7), we must have

$$c^T x^k \to Z^* \leftarrow b^T \lambda^k,$$

where Z^* is the optimal primal-dual value defined in (2.6).

In the remainder of this chapter, we show that Φ_ρ is reduced by a fixed amount $\delta > 0$, independent of n, at every step of Algorithm PR; that is,

$$\Phi_\rho(x^{k+1}, s^{k+1}) \le \Phi_\rho(x^k s^k) - \delta \qquad \text{for all } k = 0, 1, 2, \ldots . \tag{4.8}$$

The following theorem gives a general complexity result for all algorithms that achieve this reduction.

Theorem 4.4 *Given a starting point* $(x^0, \lambda^0, s^0) \in \mathcal{F}^o$*, suppose that an algorithm generates a sequence* $(x^k, \lambda^k, s^k) \in \mathcal{F}^o$ *that satisfies* (4.8) *for some* $\delta > 0$*. Then for any* $\epsilon \in (0, 1)$*, we have an index* K *defined by*

$$K = \left\lceil \frac{\Phi_\rho(x^0, s^0)}{\delta} + \frac{\rho - n}{\delta} |\log \epsilon| \right\rceil \tag{4.9}$$

such that

$$\mu_k \le \epsilon \qquad \textit{for all } k \ge K. \tag{4.10}$$

Proof. By taking the logarithm of both sides in (4.7), we find that (4.10) is a consequence of the inequality

$$\Phi_\rho(x^k, s^k) \le (\rho - n) \log \epsilon = -(\rho - n)|\log \epsilon|. \tag{4.11}$$

But from (4.8), we have

$$\Phi_\rho(x^k, s^k) \le \Phi_\rho(x^0, s^0) - k\delta, \qquad k = 1, 2, \ldots,$$

so that (4.11) is in turn implied by

$$\Phi_\rho(x^0, s^0) - k\delta \le -(\rho - n)|\log \epsilon|.$$

The result follows from a simple rearrangement of this expression. □

A Quadratic Estimate of Φ_ρ Along a Feasible Direction

We now examine a single step, in which we choose α to minimize Φ_ρ along a given search direction. In this section, we find a quadratic function $q(\alpha)$ that gives an upper bound for Φ_ρ as a function of step length α along the search direction. This approximation is valid only on a subset of the interval $(0, \alpha_{\max})$, but no matter—we can still use it to derive a conservative estimate of the reduction in Φ_ρ at this step.

We are interested mainly in search directions that satisfy the generic step equation (1.12), since these are the directions used by Algorithm PR. However, the quadratic estimate $q(\alpha)$ applies to any search direction $(\Delta x, \Delta\lambda, \Delta s)$ that maintains feasibility with respect to the equality constraints (1.4a), (1.4b); that is,

$$A\Delta x = 0, \qquad A^T\Delta\lambda + \Delta s = 0. \tag{4.12}$$

It follows immediately from (4.12) that

$$\Delta s^T \Delta x = -\Delta x^T A^T \Delta\lambda = -(A\Delta x)^T \Delta\lambda = 0. \tag{4.13}$$

We have already assumed that $\alpha \in (0, \alpha_{\max})$, so because of (4.5), the line search stays within the strictly feasible region. The quadratic estimate is valid only on the shorter interval $(0, \alpha_\tau]$ defined by

$$\alpha_\tau \max\left(\|X^{-1}\Delta x\|_\infty, \|S^{-1}\Delta s\|_\infty\right) = \tau, \tag{4.14}$$

where $\tau \in (0,1)$ is a constant to be defined later.

The quadratic estimate $q(\alpha)$ of Φ_ρ along the direction $(\Delta x, \Delta\lambda, \Delta s)$ is obtained from an application of Lemma 4.1. From (4.13) and the definition of Φ_ρ, we have

$$\begin{aligned} &\Phi_\rho(x + \alpha\Delta x, s + \alpha\Delta s) - \Phi_\rho(x, s) \\ &= \rho \log \frac{(x + \alpha\Delta x)^T(s + \alpha\Delta s)}{x^T s} - \sum_{i=1}^n \log \frac{x_i + \alpha\Delta x_i}{x_i} - \sum_{i=1}^n \log \frac{s_i + \alpha\Delta s_i}{s_i} \\ &= \rho \log \left[1 + \alpha \frac{s^T\Delta x + x^T\Delta s}{x^T s}\right] - \sum_{i=1}^n \log\left[1 + \alpha\frac{\Delta x_i}{x_i}\right] - \sum_{i=1}^n \log\left[1 + \alpha\frac{\Delta s_i}{s_i}\right]. \end{aligned} \tag{4.15}$$

Since $\alpha \in (0, \alpha_\tau] \subset (0, \alpha_{\max})$, we know that the arguments of all the logarithmic terms in (4.15) are strictly positive. By applying Lemma 4.1 to this expression (part (i) to the first term, part (ii) to the summation terms), we obtain

$$\begin{aligned} &\Phi_\rho(x + \alpha\Delta x, s + \alpha\Delta s) \\ &\le \Phi_\rho(x, s) + \rho\alpha \frac{s^T\Delta x + x^T\Delta s}{x^T s} - \alpha e^T(X^{-1}\Delta x + S^{-1}\Delta s) \\ &\qquad + \frac{\alpha^2}{2(1-\tau)}(\|X^{-1}\Delta x\|^2 + \|S^{-1}\Delta s\|^2) \\ &= \Phi_\rho(x, s) + \alpha g_1 + \tfrac{1}{2}\alpha^2 g_2 \end{aligned} \tag{4.16}$$

for all $\alpha \in (0, \alpha_\tau]$, where

$$g_1 = \rho \frac{s^T \Delta x + x^T \Delta s}{x^T s} - e^T(X^{-1}\Delta x + S^{-1}\Delta s), \tag{4.17a}$$

$$g_2 = \frac{1}{1-\tau}(\|X^{-1}\Delta x\|^2 + \|S^{-1}\Delta s\|^2). \tag{4.17b}$$

By defining the quadratic approximation as

$$q(\alpha) = \Phi_\rho(x, s) + \alpha g_1 + \tfrac{1}{2}\alpha^2 g_2, \tag{4.18}$$

we have from (4.16) that

$$\Phi_\rho(x + \alpha\Delta x, s + \alpha\Delta s) \le q(\alpha) \qquad \text{for all } \alpha \in (0, \alpha_\tau].$$

We illustrate the relationship between the functions Φ_ρ and $q(\alpha)$ in Figure 4.1. The quadratic $q(\alpha)$ is an upper bound on $\Phi_\rho(x + \alpha\Delta x, s + \alpha\Delta s)$ when $\alpha \le \alpha_\tau$. When $\alpha_{\max} < 1$, the line search approaches the boundary of the feasible set $\mathcal{F}$ as $\alpha \uparrow \alpha_{\max}$, causing the potential function Φ_ρ to approach $+\infty$. Hence, the functions $q(\alpha)$ and Φ_ρ cross over for some value of α in the interval $(\alpha_\tau, \alpha_{\max})$. When $\alpha_{\max} = 1$, on the other hand, the line search may remain in the strictly feasible set $\mathcal{F}^o$ for all $\alpha \in [0, 1]$, so Φ_ρ does not blow up, and the crossover illustrated in Figure 4.1 may not occur.

Bounding the Coefficients in the Quadratic Approximation

In this section we take a closer look at the coefficients g_1 and g_2 in the polynomial q. We restrict attention to the search direction $(\Delta x, \Delta\lambda, \Delta s)$ used in Algorithm PR, which satisfies the equation

$$S\Delta x + X\Delta s = -XSe + \frac{n}{\rho}\mu e \tag{4.19}$$

as well as the feasibility conditions (4.12). It is clear from (4.17b) that $g_2 > 0$, so q is a convex quadratic function. We will show that g_1 is negative, meaning that $q(\alpha)$ *decreases* as α increases from zero and attains a minimum at some positive value of α. By estimating the sizes of g_1 and g_2, we can estimate the steepness and deepness of the decline in q. This information, in turn, gives us an estimate of the amount of decrease in Φ_ρ, because q is an overestimate of Φ_ρ on the interval $(0, \alpha_\tau]$.

To make the analysis a little less cluttered, we define some new notation as follows. Given $(x, \lambda, s) \in \mathcal{F}^o$, we have

$$V = (XS)^{1/2}, \qquad v = Ve = \left[(x_i s_i)^{1/2}\right]_n^{i=1},$$
$$v_{\min} = \min_{i=1,\ldots,n} v_i, \qquad r = -v + \tfrac{n}{\rho}\mu V^{-1}e, \tag{4.20}$$

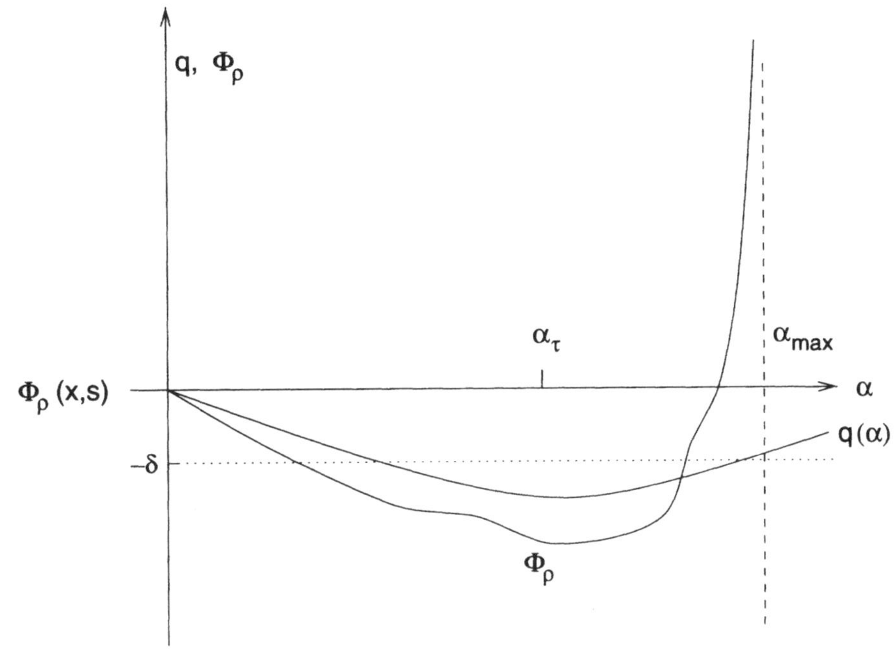

Figure 4.1. Potential function $\Phi_\rho(x + \alpha\Delta x, s + \alpha\Delta s)$ and its quadratic overestimate $q(\alpha)$.

and, as in (1.24), we define the positive diagonal matrix D by

$$D = X^{1/2}S^{-1/2}.$$

Some useful relationships between D and v and the vectors x, s include the following:

$$\|v\|^2 = x^T s = n\mu, \qquad X = VD, \qquad S = VD^{-1}. \tag{4.21}$$

The new notation can also be used to reformulate (4.19) in particularly succinct ways, including the following:

$$\begin{aligned} S\Delta x + X\Delta s &= Vr, \\ D^{-1}\Delta x + D\Delta s &= r, \\ X^{-1}\Delta x + S^{-1}\Delta s &= V^{-1}r, \end{aligned} \tag{4.22}$$

where each equation is derived from the preceding one by simply multiplying both sides by V^{-1}. We can also use the second equation in this set to derive useful bounds on various scalings of the vectors Δx and Δs. Since $\Delta x^T \Delta s = 0$ from (4.13), we have

$$\begin{aligned} \|r\|^2 &= \|D^{-1}\Delta x + D\Delta s\|^2 \\ &= \|D^{-1}\Delta x\|^2 + \Delta x^T D^{-1} D\Delta s + \|D\Delta s\|^2 \\ &= \|D^{-1}\Delta x\|^2 + \|D\Delta s\|^2, \end{aligned} \tag{4.23}$$

and therefore

$$\|D^{-1}\Delta x\| \le \|r\|, \qquad \|D\Delta s\| \le \|r\|. \tag{4.24}$$

We can push this line of reasoning a little further to obtain a bound on the coefficient g_2 of the quadratic term in q. From (4.21) and (4.23), we have

$$\begin{aligned} \|X^{-1}\Delta x\|^2 + \|S^{-1}\Delta s\|^2 &= \|V^{-1}D^{-1}\Delta x\|^2 + \|V^{-1}D\Delta s\|^2 \\ &\le \|V^{-1}\|^2 \left(\|D^{-1}\Delta x\|^2 + \|D\Delta s\|^2\right) \\ &\le \frac{1}{v_{\min}^2}\|r\|^2. \end{aligned} \tag{4.25}$$

Hence, from the definition of g_2 (4.17b), we have

$$g_2 \le \frac{1}{1-\tau}\frac{\|r\|^2}{v_{\min}^2}. \tag{4.26}$$

The linear coefficient g_1 in q can also be expressed in terms of the vector r. By using (4.22) and the expression for r in (4.20), we obtain

$$\begin{aligned} g_1 &= \frac{\rho}{x^T s} e^T(S\Delta x + X\Delta s) - e^T(X^{-1}\Delta x + S^{-1}\Delta s) \\ &= \frac{\rho}{x^T s} e^T V r - e^T V^{-1} r \\ &= -\frac{\rho}{n\mu}\left[-v + \frac{n}{\rho}\mu V^{-1} e\right]^T r \\ &= -\frac{\rho}{n\mu}\|r\|^2, \end{aligned} \tag{4.27}$$

where the last equality follows from (4.20).

The formulas (4.25) and (4.27) give us estimates for both g_1 and g_2 in terms of $\|r\|^2$. Assuming that $r \ne 0$—a fact proven below—we have that $g_1 < 0$. Since $g_2 > 0$, the graph of q is an upturned parabola, as shown in Figure 4.1.

An Estimate of the Reduction in Φ_ρ and Polynomial Complexity

We now guarantee a significant decrease in Φ_ρ along the search direction by showing that the minimum value of $q(\alpha)$ in the range $(0, \alpha_\tau]$ is significantly smaller than $q(0)$. By proving that $\|r\|$ is not too small, we deduce that g_1 is negative enough to ensure that $q(\alpha)$ takes a sharp dip as α increases from zero.

Lemma 4.5 obtains a lower bound on $\|r\|$, and Theorem 4.6 uses this bound to estimate the improvement in q and Φ_ρ.

Lemma 4.5 *For any $(x, \lambda, s) \in \mathcal{F}^o$ and for $\rho > n + \sqrt{n}$, we have*

$$\|r\| \geq \frac{\sqrt{3}}{2v_{\min}} \frac{n\mu}{\rho}. \tag{4.28}$$

Proof. By using the definition (4.20), we have

$$\begin{aligned}
\frac{\rho^2}{n^2\mu^2}\|r\|^2 &= \left\| \frac{\rho}{n\mu} v - V^{-1}e \right\|^2 \\
&= (V^{-1}e)^T(V^{-1}e) - 2\frac{\rho}{n\mu} v^T V^{-1}e + \frac{\rho^2}{n^2\mu^2} v^T v \\
&= (V^{-1}e)^T(V^{-1}e) - 2\frac{\rho}{\mu} + \frac{\rho^2}{n\mu} \\
&= (V^{-1}e)^T(V^{-1}e) + \frac{\rho^2 - 2n\rho + n^2 - n}{n\mu} - \frac{n^2 - n}{n\mu},
\end{aligned} \tag{4.29}$$

where we obtained the last line by simply adding and subtracting the term $(n^2 - n)/(n\mu)$. The middle term in this last expression is nonnegative for $\rho \geq n + \sqrt{n}$ and zero for $\rho = n + \sqrt{n}$, since

$$\rho^2 - 2n\rho + n^2 - n = (\rho - n - \sqrt{n})^2 + 2\sqrt{n}(\rho - n - \sqrt{n}).$$

We therefore have

$$\frac{\rho^2}{n^2\mu^2}\|r\|^2 \geq (V^{-1}e)^T(V^{-1}e) - \frac{n^2 - n}{n\mu}, \tag{4.30}$$

with equality if $\rho = n + \sqrt{n}$.

To find a lower bound on this last expression, we use the observation that v is orthogonal to $V^{-1}e - v/\mu$ (see the exercises). By substituting $\rho = n + \sqrt{n}$

into (4.30) and using the definition of r from (4.20), we obtain

$$
\begin{aligned}
(V^{-1}e)^T(V^{-1}e) - \frac{n^2-n}{n\mu} &= \left\| V^{-1}e - \frac{n+\sqrt{n}}{n\mu}v \right\|^2 \\
&= \left\| V^{-1}e - \frac{1}{\mu}v \right\|^2 + \left\| \frac{\sqrt{n}}{n\mu}v \right\|^2 .
\end{aligned}
$$

The first term can be bounded below by the square of any one component of the vector $V^{-1}e - v/\mu$; we choose the component that achieves $v_{\min}$ to obtain

$$
\left\| V^{-1}e - \frac{1}{\mu}v \right\| \geq \left| \frac{1}{v_{\min}} - \frac{1}{\mu}v_{\min} \right| .
$$

Using (4.21), (4.30), and simple manipulation, we obtain

$$
\begin{aligned}
\frac{\rho^2}{n^2\mu^2}\|r\|^2 &\geq \left[\frac{1}{v_{\min}} - \frac{1}{\mu}v_{\min} \right]^2 + \frac{1}{n\mu^2}v^Tv \\
&= \frac{1}{\mu^2}\left\{ \left[\frac{\mu}{v_{\min}} - v_{\min} \right]^2 + \mu \right\} \\
&= \frac{1}{\mu^2}\left\{ \left[\frac{\mu}{2v_{\min}} - v_{\min} \right]^2 + \frac{3}{4}\frac{\mu^2}{v_{\min}^2} \right\} \\
&\geq \frac{3}{4v_{\min}^2} .
\end{aligned}
$$

The result (4.28) follows immediately. □

To complete the analysis, we find a value of α that achieves a fixed reduction in $q(\alpha)$ and Φ_ρ. For simplicity, we will work with the specific value $\tau = 0.5$. Similar results can be proven for other values of τ in the range $(0, 1)$.

Theorem 4.6 *Let $\tau = 0.5$, and define*

$$
\bar{\alpha} = \frac{v_{\min}}{2\|r\|} . \tag{4.31}
$$

Then

$$
q(\bar{\alpha}) - q(0) \leq -0.15,
$$

.;./, and so (4.8) holds with $\delta = 0.15$.

Proof. We check first that $\bar{\alpha}$ satisfies the restriction (4.14). Since from (4.25) we have

$$\|X^{-1}\Delta x\|_2 \leq \frac{\|r\|}{v_{\min}},$$

it follows that

$$\bar{\alpha}\|X^{-1}\Delta x\|_\infty \leq \bar{\alpha}\|X^{-1}\Delta x\|_2 \leq \tfrac{1}{2}\frac{v_{\min}}{\|r\|}\frac{\|r\|}{v_{\min}} = \tfrac{1}{2} = \tau.$$

The other bound in (4.14) can be verified in the same way.

From the estimates (4.26) and (4.27) of g_2 and g_1, we have by substituting $\bar{\alpha}$ from (4.31) that

$$\begin{aligned} q(\bar{\alpha}) - q(0) &= \bar{\alpha} g_1 + \tfrac{1}{2}\bar{\alpha}^2 g_2 \\ &\leq -\frac{\rho}{n\mu}\bar{\alpha}\|r\|^2 + \tfrac{1}{2}\bar{\alpha}^2 \frac{1}{v_{\min}^2}\frac{1}{1-\tau}\|r\|^2 \\ &= -\frac{\rho}{n\mu}\frac{v_{\min}}{2}\|r\| + \frac{1}{8(1-\tau)}. \end{aligned}$$

Using the lower bound (4.28) on $\|r\|$ and substituting $\tau = \frac{1}{2}$, we obtain

$$q(\bar{\alpha}) - q(0) \leq -\frac{\sqrt{3}}{4} + \frac{1}{8(1-.5)} = \frac{1-\sqrt{3}}{4} \leq -0.15.$$

Finally, from (4.16) and the definition (4.18), we find that

$$\begin{aligned} &\min_{\alpha \in (0,\alpha_{\max})} \Phi_\rho(x + \alpha\Delta x, s + \alpha\Delta s) \\ &\leq \Phi_\rho(x + \bar{\alpha}\Delta x, s + \bar{\alpha}\Delta s) \leq q(\bar{\alpha}) \leq q(0) - 0.15 = \Phi_\rho(x,s) - 0.15. \end{aligned}$$

Hence, (4.8) holds with $\delta = 0.15$, as claimed. □

The actual value of δ is not too important. The analysis is somewhat pessimistic, so the reductions obtained in practice are usually larger than this lower bound.

We wrap up the analysis with a formal complexity result based on Theorem 4.4.

Corollary 4.7 *Fix $\rho \geq n + \sqrt{n}$ and $\epsilon > 0$, and suppose that our initial point $(x^0, \lambda^0, s^0) \in \mathcal{F}^o$ has*

$$\Phi_\rho(x^0, s^0) \leq \kappa(\rho - n)|\log \epsilon|$$

for some $\kappa > 0$ independent of n. Then the index K defined by

$$K = \left\lceil \frac{\kappa + 1}{0.15} (\rho - n) |\log \epsilon| \right\rceil = O\left((\rho - n)|\log \epsilon|\right) \tag{4.32}$$

has the property that

$$(x^k, \lambda^k, s^k) \in \mathcal{F}^o, \qquad \mu_k \leq \epsilon,$$

for all $k \geq K$.

This result is easy to verify by substituting the bound on $\Phi_\rho(x^0, s^0)$ into the expression (4.9).

What About Centrality?

Potential-reduction algorithms generate the same generic search directions (1.12) as path-following algorithms, but they make no explicit requirement that the iterates (x, λ, s) stay in a neighborhood of the central path. Centrality still plays an implicit role, however. In particular, the step length $\bar{\alpha}$ tends to be shorter if (x, λ, s) is poorly centered. By substituting (4.28) into (4.31), we find that

$$\bar{\alpha} \leq \tfrac{1}{2} v_{\min} \frac{2 v_{\min}}{\sqrt{3}} \frac{\rho}{n\mu} = \frac{\rho}{\sqrt{3}n} \frac{v_{\min}^2}{\mu}. \tag{4.33}$$

A perfectly centered point has $x_i s_i = \mu$ for all i, so that $v_{\min}^2 = \mu$. By contrast, poorly centered points have $v_{\min}^2 \ll \mu$, leading to an upper bound

$$\alpha \ll \frac{\rho}{\sqrt{3}n}.$$

If we make the optimal choice $\rho = n + \sqrt{n}$, this bound suggests $\alpha \ll 1$.

Larger values of ρ do not help much. For instance, if $\rho > 4n$, then (4.29) yields

$$\frac{\rho^2}{n^2\mu^2} \|r\|^2 = (V^{-1}e)^T (V^{-1}e) - \frac{2\rho}{\mu} + \frac{\rho^2}{n\mu} \geq v_{\min}^{-2} + \frac{\rho^2}{2n\mu} \geq \frac{\rho^2}{2n\mu}$$

so that substitution in (4.31) yields

$$\alpha = \frac{v_{\min}}{2\|r\|} \leq \frac{1}{\sqrt{2n}} \frac{v_{\min}}{\sqrt{\mu}}.$$

Again, poor centrality implies that $v_{\min} \ll \sqrt{\mu}$ and therefore $\alpha \ll 1$.

Given that Algorithm PR decreases Φ_ρ monotonically to $-\infty$, the summation term $\sum -\log(x_i s_i/\mu)$ prevents the iterates from becoming *too* poorly centered, at least until μ becomes small. If $x_j s_j \ll \mu$ for some j, then the j term in this summation contributes a large positive quantity to Φ_ρ, which can be balanced only if the contribution from the first term $(\rho - n)\log x^T s$ is large (that is, μ is small). We can be a little more specific about this observation. Given any small positive number $\bar{\epsilon} \in (0, 1]$, the iterates of Algorithm PR eventually enter the level set $\mathcal{L}$ defined by

$$\mathcal{L} = \{(x, \lambda, s) \in \mathcal{F}^o \mid \Phi_\rho(x, s) \le \log \bar{\epsilon} + (n-1)\log n - 1\}.$$

For any index $j = 1, 2, \ldots, n$, we find that

$$\Phi_\rho(x, s) \ge (n-1)\log n - 1 - \log \frac{x_j s_j}{(x^T s)^{\rho-n+1}} \tag{4.34}$$

(see the exercises), so for points $(x, \lambda, s) \in \mathcal{L}$, we have

$$-\log \frac{x_j s_j}{(x^T s)^{\rho-n+1}} \le \log \bar{\epsilon} \;\Rightarrow\; x_j s_j \ge \frac{(x^T s)^{\rho-n}}{\bar{\epsilon}} x^T s. \tag{4.35}$$

Therefore, each pairwise product $x_j s_j$ is bounded away from zero by a multiple of some positive power of μ. We illustrate the neighborhood $\mathcal{L}$ for the case of $n = 2$ in Figure 4.2. We plot $x_1 s_1$ against $x_2 s_2$ to highlight centrality; points too close to either of the axes are poorly centered, since $x_1 s_1$ and $x_2 s_2$ are very different. Similar figures appear in Chapter 5 to illustrate the neighborhoods used by path-following algorithms (see Figures 5.1, 5.2, and so on).

Choosing ρ and α

Our complexity result, Corollary 4.7, holds for all values of ρ with $\rho \ge n + \sqrt{n}$. In the first complete analysis of Algorithm PR—the paper of Kojima, Mizuno, and Yoshise [68]—ρ was fixed at its lower bound $n + \sqrt{n}$. This choice yields the best complexity estimate—$O(\sqrt{n}|\log \epsilon|)$ in (4.32)—but it leads to a conservative choice of centering parameter σ, namely,

$$\sigma = \frac{n}{\rho} = \frac{n}{n + \sqrt{n}} \approx 1 - \frac{1}{\sqrt{n}} \qquad \text{for large } n.$$

Smaller values of σ can be obtained by choosing larger values of ρ—for example, $\rho = 10n$ or $\rho = n + n^{1.5}$. These values degrade the complexity estimate to $O(n|\log \epsilon|)$ and $O(n^{1.5}|\log \epsilon|)$, respectively, but they usually

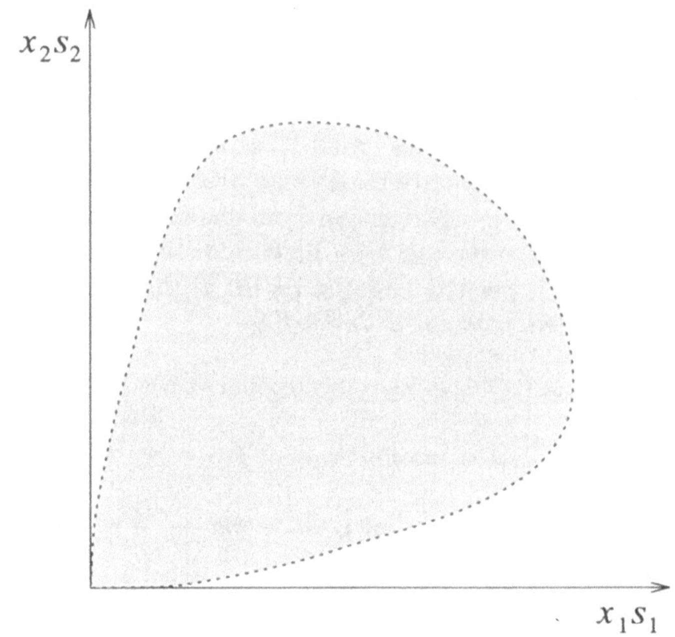

Figure 4.2. Level set of $\Phi_\rho(x, s)$ for $n = 2$, plotted in the (xs) space.

produce better numerical performance.

The minimizing value of the step length α used in Algorithm PR is appealing from the viewpoint of the analysis but is not really practical. First, since Φ_ρ is a highly nonlinear function, it is not convenient to find the *exact* minimizing value of α. This difficulty is easy to circumvent—an approximate minimizer is good enough for the analysis to proceed. Second, the minimizing α often is too short to produce reasonable computational performance. Values such as $0.99\alpha_{\max}$ and $0.95\alpha_{\max}$ tend to do better in practice. It is easy to derive heuristics that compromise between theory and practice in choosing α. For instance, we could first try a few larger values of α, such as $\alpha = 0.999\alpha_{\max}$, $\alpha = 0.99\alpha_{\max}$, and select one of these values as the step if it produces at least a prescribed constant decrease in Φ_ρ. If all the larger values fail, we could fall back on an approximate line search, or even the easily computed value $\bar{\alpha}$ from (4.31).

Notes and References

The results in this chapter were drawn principally from Kojima, Mizuno, and Yoshise [68]. Similar lemmas were proven by other authors, notably Ye [154]. Some results, such as Lemma 4.3, were drawn from Todd's survey paper [128], which also discusses primal potential-reduction methods and outlines current research on extensions to certain convex nonlinear problems. In their monograph [65], Kojima et al. discussed connections with path-following methods.

Infeasible-interior-point algorithms based on the method of this chapter have been described by Mizuno, Kojima, and Todd [91], but they are not quite as elegant as the infeasible-path-following method discussed in Chapter 6.

Exercises

1. Prove part (i) of Lemma 4.1.

2. Prove that for $\alpha_{\max}$ defined by (4.5) and α_τ defined in (4.14), we have $0 < \alpha_\tau < \alpha_{\max}$.

3. Show that v is orthogonal to $V^{-1}e - v/\mu$, where V and v are defined in (4.20).

4. Let $(\Delta x, \Delta\lambda, \Delta s)$ be the search direction in Algorithm PR, and let $\alpha_{\max}$ be defined as in (4.5). Show that $\log((x + \alpha\Delta x)^T(s + \alpha\Delta s))$ is a concave function of α for $\alpha \in (0, \alpha_{\max})$.

5. By avoiding the specific choice $\tau = 0.5$ in Theorem 4.6 and setting $\bar{\alpha} = \tau v_{\min}/\|r\|$, find an estimate for $q(\bar{\alpha}) - q(0)$ as a function of τ.

6. Verify the inequality (4.34). Hint: Write the function $\Phi_\rho(x, s)$ as

$$\Phi_\rho(x, s) = -\sum_{i \neq j} \log \frac{x_i s_i}{x^T s} - \log \frac{x_j s_j}{(x^T s)^{\rho - n + 1}},$$

   and use the technique of Lemma 4.2 to obtain a lower bound on the first term on the right-hand side.

Chapter 5

Path-Following Algorithms

In Chapters 1 and 2, we described the central path $\mathcal{C}$, a path of points $(x_\tau, \lambda_\tau, s_\tau)$ that leads to the set Ω of primal-dual solutions. Points in Ω satisfy the KKT conditions (2.4), whereas points in $\mathcal{C}$ are defined by conditions that differ from the KKT conditions only by the presence of a positive scalar parameter $\tau > 0$, namely,

$$\begin{aligned} A^T\lambda + s &= c, && && (5.1a) \\ Ax &= b, && && (5.1b) \\ (x, s) &\geq 0, && && (5.1c) \\ x_i s_i &= \tau, && i = 1, 2, \ldots, n. && (5.1d) \end{aligned}$$

We showed in Chapter 2 that this system has a unique solution $(x_\tau, \lambda_\tau, s_\tau)$ for each $\tau > 0$ whenever the problem is feasible (although the KKT conditions, for which $\tau = 0$ in (5.1d), may have multiple solutions).

Path-following methods follow $\mathcal{C}$ in the direction of decreasing τ to the solution set Ω. They do not necessarily stay exactly *on* $\mathcal{C}$ or even particularly close to it. Rather, they stay within a loose but well-defined neighborhood of $\mathcal{C}$ while steadily reducing the duality measure μ to zero. Each search direction is a Newton step toward a point on $\mathcal{C}$, a point for which the duality measure τ is equal to or smaller than the current duality measure μ. The target value $\tau = \sigma\mu$ is used, where $\sigma \in [0, 1]$ is the centering parameter introduced in Chapter 1.

The algorithms of this chapter generate strictly feasible iterates (x^k, λ^k, s^k) that satisfy the first three KKT conditions (5.1a), (5.1b), and (5.1c). They deviate from the central path $\mathcal{C}$ only because the pairwise products $x_i s_i$ are generally not identical, so the condition (5.1d) is not satisfied exactly. This deviation is measured by comparing the pairwise products with their average value $\mu = x^T s/n = (\sum x_i s_i)/n$, using, for example, a scaled norm defined by

$$\frac{1}{\mu}\|XSe - \mu e\| = \frac{1}{\mu}\left\| \begin{bmatrix} x_1 s_1 \\ \vdots \\ x_n s_n \end{bmatrix} - \left(\frac{x^T s}{n}\right) e \right\|. \tag{5.2}$$

In the literature, both the 2-norm and the ∞-norm have been used in this definition. For both norms, we can ensure that x and s are *strictly* positive by requiring that $(1/\mu)\|XSe - \mu e\| < 1$. (If component i of x or s is zero, we have $\|XSe - \mu e\| \geq |x_i s_i - \mu| = \mu$.)

By using the 2-norm in (5.2) and restricting the deviation to be less than a constant $\theta \in [0, 1)$, we obtain the neighborhood $\mathcal{N}_2(\theta)$ defined in (1.15):

$$\mathcal{N}_2(\theta) = \{(x, \lambda, s) \in \mathcal{F}^o \mid \|XSe - \mu e\|_2 \leq \theta\mu\}. \tag{5.3}$$

By using the ∞-norm in (5.2), we obtain the neighborhood $\mathcal{N}_\infty(\theta)$. We can motivate another neighborhood $\mathcal{N}_{-\infty}(\gamma)$ by noting that our chief concern is to keep the products $x_i s_i$ from becoming too much *smaller* than their average value μ and therefore to prevent x and s from approaching the boundary of the region $(x, s) \geq 0$ prematurely. We do not mind if some of these products are somewhat *larger* than μ, so the neighborhood $\mathcal{N}_{-\infty}(\gamma)$ uses a one-sided bound on $x_i s_i$ in place of the two-sided bound in (5.2), that is,

$$\mathcal{N}_{-\infty}(\gamma) = \{(x, \lambda, s) \in \mathcal{F}^o \mid x_i s_i \geq \gamma\mu \quad \text{for all } i = 1, 2, \ldots, n\}, \tag{5.4}$$

where $\gamma \in (0, 1)$.

Path-following methods follow Framework PD of Chapter 1. They select one of the neighborhood types $\mathcal{N}_2$, $\mathcal{N}_\infty$, or $\mathcal{N}_{-\infty}$ and choose the centering parameter σ and the step length parameter α to ensure that every iterate (x^k, λ^k, s^k) stays within the chosen neighborhood.

Methods based on the neighborhood $\mathcal{N}_2$ have $O(\sqrt{n} \log 1/\epsilon)$ complexity, matching the complexity estimate for the potential-reduction method of Chapter 4 (see Corollary 4.7). We describe two such methods in this chapter: the short-step path-following algorithm (Algorithm SPF) and the predictor-corrector algorithm (Algorithm PC). Algorithm SPF, the simplest of all interior-point methods, chooses a constant value $\sigma_k \equiv \sigma$ for the centering parameter and fixes the step length at $\alpha_k \equiv 1$ for all iterations k. This method was introduced by Kojima, Mizuno, and Yoshise [66] and Monteiro and Adler [94]; our analysis follows the latter paper. Algorithm PC alternates between two types of steps: *predictor* steps, which improve the value of μ but which also tend to worsen the centrality measure (5.2), and *corrector* steps, which have no effect on the duality measure μ but improve centrality. Various aspects of this algorithm were foreshadowed by a number of authors, including Monteiro and Adler [94] and Sonnevend, Stoer, and Zhao [122, 123], but it was first stated and analyzed in the simple form used here by Mizuno, Todd, and Ye [92]. The algorithm is sometimes called "Mizuno–Todd–Ye predictor-corrector" to distinguish it from the quite different Mehrotra predictor-corrector algorithm of Chapter 10.

A disadvantage of the $\mathcal{N}_2(\theta)$ neighborhood is its restrictive nature. From the definition (5.3), we have for $(x, \lambda, s) \in \mathcal{N}_2(\theta)$ that

$$\sum_{i=1}^{n} \left(\frac{x_i s_i}{\mu} - 1 \right)^2 \leq \theta^2 < 1$$

so that the sum of squares of *all* relative deviations of $x_i s_i$ from their average value μ cannot exceed 1. Even if θ is close to its upper bound of 1, the neighborhood $\mathcal{N}_2(\theta)$ contains only a small fraction of the points in the strictly feasible set $\mathcal{F}^o$, so algorithms based on this neighborhood do not have much room in which to maneuver and the amount of progress they can achieve at each iteration is limited. The neighborhood $\mathcal{N}_{-\infty}(\gamma)$, on the other hand, is much more expansive: When γ is small, it takes up almost the entire strictly feasible set $\mathcal{F}^o$. We discuss a long-step path-following algorithm based on this neighborhood—Algorithm LPF—that makes more aggressive (that is, smaller) choices of centering parameter σ than does Algorithm SPF. Instead of taking unit steps, however, Algorithm LPF performs a line search along the direction obtained from (1.13), choosing α_k to be as large as possible subject to the restriction of remaining within $\mathcal{N}_{-\infty}(\gamma)$. Algorithm LPF is closely related to the very first polynomial primal-dual algorithm proposed by Kojima, Mizuno, and Yoshise [67]. It is more closely related to practical algorithms than are Algorithms SPF and PC, but its complexity bound is worse: $O(n \log 1/\epsilon)$ vs. $O(\sqrt{n} \log 1/\epsilon)$.

Although the kth step taken by a path-following algorithm aims for the point on the central path $\mathcal{C}$ whose duality measure is $\sigma_k \mu_k$, it rarely hits this target. The reason is that there is a discrepancy between the nonlinear equations (5.1) and the linear approximation on which the Newton-like step equations (1.13) are based. This discrepancy is quantified by the pairwise products $\Delta x_i \Delta s_i$. Much of the analysis in this chapter is concerned with finding bounds on these products and showing that the step $(\Delta x^k, \Delta \lambda^k, \Delta s^k)$ makes significant progress toward its target without actually scoring a direct hit.

In this chapter, we focus mainly on the convergence of the sequence $\{\mu_k\}$ of duality measures to zero. We close the chapter by looking at a different, but related, issue: convergence of the primal-dual iteration sequences $\{(x^k, \lambda^k, s^k)\}$. We prove that the x^k and s^k components are bounded and that strictly complementary solutions to (2.1), (2.2) can be recovered from the limit points of these sequences.

All three algorithms described in this chapter are remarkable for their simplicity. Despite their strong theoretical properties, they are easy to state and to analyze, once we are past the hurdle of a few tricky technical results. They also provide the foundation for more powerful algorithms, including algorithms that allow infeasible starting points and rapid local convergence. Unfortunately, as we see in subsequent chapters, our claim of simple analysis does not necessarily hold when these capabilities are added.

The Short-Step Path-Following Algorithm

We start with the short-step path-following algorithm, Algorithm SPF. This method starts at a point $(x^0, \lambda^0, s^0) \in \mathcal{N}_2(\theta)$ and uses uniform values $\alpha_k = 1$ and $\sigma_k = \sigma$, where θ and σ satisfy a certain relationship, described below. All iterates (x^k, λ^k, s^k) stay inside $\mathcal{N}_2(\theta)$, and the duality measure μ_k converges linearly to zero at the constant rate $1 - \sigma$.

The method is defined by filling in the Framework PD from Chapter 1. We assign specific values to θ and σ, justifying them in the analysis that follows.

Algorithm SPF
Given $\theta = 0.4$, $\sigma = 1 - 0.4/\sqrt{n}$, and $(x^0, \lambda^0, s^0) \in \mathcal{N}_2(\theta)$;
for $k = 0, 1, 2, \ldots$
  set $\sigma_k = \sigma$ and solve (1.13) to obtain $(\Delta x^k, \Delta\lambda^k, \Delta s^k)$;
  set $(x^{k+1}, \lambda^{k+1}, s^{k+1}) = (x^k, \lambda^k, s^k) + (\Delta x^k, \Delta\lambda^k, \Delta s^k)$;
end (for).

This algorithm is illustrated in Figure 5.1, which plots the first few iterates of the algorithm projected into an unusual space. The horizontal and vertical axes represent the pairwise products $x_1 s_1$ and $x_2 s_2$ for this two-dimensional problem, so the central path is the line emanating from $(0, 0)$ at an angle of $\pi/4$ radians. In this (nonlinear) space, the search directions transform to curves rather than straight lines. The solution is at the origin, and the challenge facing the algorithm is to reach this point while maintaining the feasibility conditions $Ax = b$, $A^T\lambda + s = c$ at all iterates.

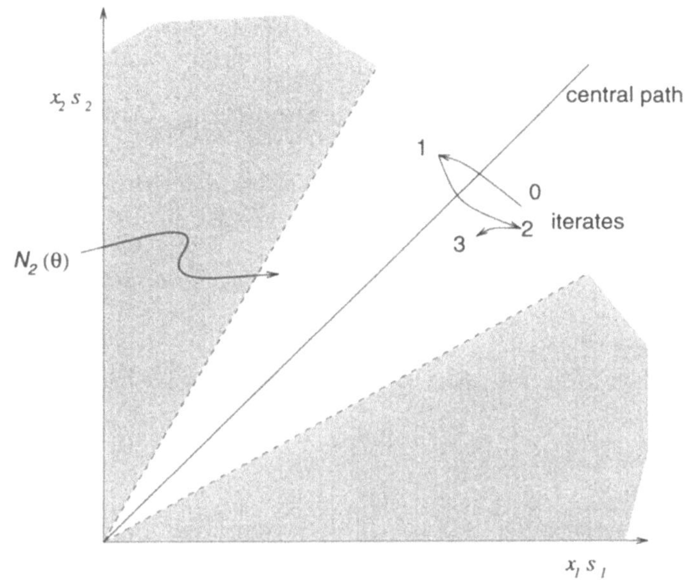

Figure 5.1. Iterates of Algorithm SPF, plotted in (xs) space.

The representations of the central path $\mathcal{C}$ and the neighborhoods $\mathcal{N}_2(\theta)$ and $\mathcal{N}_{-\infty}(\gamma)$ in Figures 5.1, 5.2, and 5.3 make it easy to see how step lengths are affected by neighborhood boundaries.

Most of the effort in the analysis of Algorithm SPF goes into showing that all its iterates stay in the neighborhood $\mathcal{N}_2(\theta)$. We prove this claim below in Lemmas 5.4 and 5.5 and Theorem 5.6. For now, let us take this claim on trust and prove the global convergence and polynomial complexity results. Linear convergence follows from Lemma 5.1, which we state in general terms because it applies to *all* algorithms that obtain their search directions from the system (1.12). The following notation is useful:

$$(x(\alpha), \lambda(\alpha), s(\alpha)) = (x, \lambda, s) + \alpha(\Delta x, \Delta\lambda, \Delta s), \tag{5.5a}$$

$$\mu(\alpha) = x(\alpha)^T s(\alpha)/n. \tag{5.5b}$$

Lemma 5.1 *Let the step* $(\Delta x, \Delta\lambda, \Delta s)$ *be defined by* (1.12). *Then*

$$\Delta x^T \Delta s = 0 \tag{5.6}$$

and

$$\mu(\alpha) = (1 - \alpha(1 - \sigma))\mu. \tag{5.7}$$

Proof. The first result is left as an exercise. For the second result, we use the third row of (1.12), namely,

$$S\Delta x + X\Delta s = -XSe + \sigma\mu e. \tag{5.8}$$

By summing the n components of this equation, we obtain $s^T\Delta x + x^T\Delta s = -(1-\sigma)x^Ts$. From this formula and (5.6), we obtain

$$x(\alpha)^T s(\alpha) = x^T s + \alpha(s^T\Delta x + x^T\Delta s) + \alpha^2 \Delta x^T \Delta s = x^T s(1-\alpha(1-\sigma)),$$

giving the result. □

For Algorithm SPF, we have by our specific choices of σ_k and α_k that

$$\mu_{k+1} = \sigma\mu_k = \left(1 - \frac{0.4}{\sqrt{n}}\right)\mu_k, \qquad k = 0, 1, \ldots, \tag{5.9}$$

so global linear convergence follows. The polynomial complexity result is an immediate consequence of (5.9) and Theorem 3.2.

Theorem 5.2 *Given $\epsilon > 0$, suppose that the starting point $(x^0, \lambda^0, s^0) \in \mathcal{N}_2(0.4)$ in Algorithm SPF has*

$$\mu_0 \le 1/\epsilon^\kappa$$

for some positive constant κ. Then there is an index K with $K = O(\sqrt{n}\log 1/\epsilon)$ such that

$$\mu_k \le \epsilon \qquad \textit{for all } k \ge K.$$

Proof. Because of (5.9), we can set $\delta = 0.4$ and $\omega = 0.5$ in Theorem 3.2, and the result follows immediately. □

In the next section, we return to the task of showing that the iterates (x^k, λ^k, s^k) stay inside $\mathcal{N}_2(\theta)$.

Technical Results

The first result is purely technical; its proof can be found at the end of the chapter.

Lemma 5.3 *Let u and v be any two vectors in $\mathbb{R}^n$ with $u^Tv \ge 0$. Then*

$$\|UVe\| \le 2^{-3/2}\|u+v\|^2,$$

where

$$U = \mathrm{diag}(u_1, u_2, \ldots, u_n), \qquad V = \mathrm{diag}(v_1, v_2, \ldots, v_n).$$

We showed in Lemma 5.1 that the inner product $\Delta x^T \Delta s = \sum \Delta x_i \Delta s_i$ is zero, but there is no reason for the individual pairwise products to be zero. In the next result, Lemma 5.4, we find a bound on the vector of these pairwise products.

Lemma 5.4 *If* $(x, \lambda, s) \in \mathcal{N}_2(\theta)$, *then*

$$\|\Delta X \Delta S e\| \le \frac{\theta^2 + n(1-\sigma)^2}{2^{3/2}(1-\theta)}\mu.$$

Proof. Recall the definition of the diagonal matrix D as $X^{1/2}S^{-1/2}$ from (1.24). If we multiply (5.8) by $(XS)^{-1/2}$, we obtain

$$D^{-1}\Delta x + D\Delta s = (XS)^{-1/2}(-XSe + \sigma\mu e). \tag{5.10}$$

Now, apply Lemma 5.3 with $u = D^{-1}\Delta x$ and $v = D\Delta s$ to obtain

$$\begin{aligned}
\|\Delta X \Delta S e\| &= \|(D^{-1}\Delta X)(D\Delta S)e\| \\
&\le 2^{-3/2}\|D^{-1}\Delta x + D\Delta s\|^2 && \text{from Lemma 5.3} \\
&= 2^{-3/2}\|(XS)^{-1/2}(-XSe + \sigma\mu e)\|^2 && \text{from (5.10)} \\
&= 2^{-3/2}\sum_{i=1}^{n}\frac{(-x_i s_i + \sigma\mu)^2}{x_i s_i} \\
&\le 2^{-3/2}\frac{\|XSe - \sigma\mu e\|^2}{\min_i x_i s_i}. && (5.11)
\end{aligned}$$

Since $(x, \lambda, s) \in \mathcal{N}_2(\theta)$, we have

$$\min_i x_i s_i \ge (1-\theta)\mu. \tag{5.12}$$

For a bound on the numerator in (5.11), note first that

$$e^T(XSe - \mu e) = x^T s - \mu e^T e = 0,$$

and therefore

$$\begin{aligned}
&\|XSe - \sigma\mu e\|^2 \\
&\quad= \|(XSe - \mu e) + (1-\sigma)\mu e\|^2 \\
&\quad= \|XSe - \mu e\|^2 + 2(1-\sigma)\mu e^T(XSe - \mu e) + (1-\sigma)^2\mu^2 e^T e \\
&\quad\le \theta^2\mu^2 + (1-\sigma)^2\mu^2 n. && (5.13)
\end{aligned}$$

We obtain the result by substituting (5.12) and (5.13) into (5.11). □

Lemma 5.10 is a result similar to Lemma 5.4 that is proven during our analysis of Algorithm LPF. Lemma 5.4 gives a tighter bound, however, because the current point (x, λ, s) lies in the more restrictive neighborhood $\mathcal{N}_2(\theta)$. The difference in the bounds obtained in Lemmas 5.4 and 5.10 accounts for the difference in polynomial complexity estimate for Algorithms SPF and LPF.

We know from Lemma 5.1 that μ decreases linearly as we move along the direction $(\Delta x, \Delta\lambda, \Delta s)$. But how far does the point $(x(\alpha), \lambda(\alpha), s(\alpha))$ stray from the central path in the 2-norm measure? The answer is provided by the following result, which is a simple consequence of Lemma 5.4.

Lemma 5.5 *If* $(x, \lambda, s) \in \mathcal{N}_2(\theta)$, *we have*

$$
\begin{aligned}
&\|X(\alpha)S(\alpha)e - \mu(\alpha)e\| \\
&\quad \leq \ |1-\alpha|\,\|XSe - \mu e\| + \alpha^2\|\Delta X \Delta S e\| &\text{(5.14a)}\\
&\quad \leq \ |1-\alpha|\theta\mu + \alpha^2\left[\frac{\theta^2 + n(1-\sigma)^2}{2^{3/2}(1-\theta)}\right]\mu. &\text{(5.14b)}
\end{aligned}
$$

Proof. We use the third row of (1.12) to resolve the components of $X(\alpha)S(\alpha)e - \mu(\alpha)e$. From this equation and Lemma 5.1, we have

$$
\begin{aligned}
&x_i(\alpha)s_i(\alpha) - \mu(\alpha) \\
&\quad = \ x_is_i + \alpha(s_i\Delta x_i + x_i\Delta s_i) + \alpha^2\Delta x_i\Delta s_i - (1-\alpha(1-\sigma))\mu \\
&\quad = \ x_is_i(1-\alpha) + \alpha\sigma\mu + \alpha^2\Delta x_i\Delta s_i - (1-\alpha+\alpha\sigma)\mu \\
&\quad = \ x_is_i(1-\alpha) + \alpha^2\Delta x_i\Delta s_i - (1-\alpha)\mu.
\end{aligned}
$$

Reassembling these components into a vector, we obtain

$$
\begin{aligned}
&\|X(\alpha)S(\alpha)e - \mu(\alpha)e\| \\
&\quad = \ \left\|\left[x_is_i(1-\alpha) - (1-\alpha)\mu + \alpha^2\Delta x_i\Delta s_i\right]_{i=1}^n\right\| \\
&\quad \leq \ |1-\alpha|\,\|XSe - \mu e\| + \alpha^2\|\Delta X\Delta S e\| \\
&\quad \leq \ |1-\alpha|\,\theta\mu + \alpha^2\left[\frac{\theta^2 + n(1-\sigma)^2}{2^{3/2}(1-\theta)}\right]\mu. \quad \square
\end{aligned}
$$

Theorem 5.6 defines a relationship between θ and σ and shows that even a full step ($\alpha = 1$) along the search direction will not take the new iterate outside the neighborhood $\mathcal{N}_2(\theta)$.

Theorem 5.6 *Let the parameters $\theta \in (0,1)$ and $\sigma \in (0,1)$ be chosen to satisfy*

$$\frac{\theta^2 + n(1-\sigma)^2}{2^{3/2}(1-\theta)} \leq \sigma\theta. \tag{5.15}$$

Then if $(x, \lambda, s) \in \mathcal{N}_2(\theta)$, we have

$$(x(\alpha), \lambda(\alpha), s(\alpha)) \in \mathcal{N}_2(\theta)$$

for all $\alpha \in [0,1]$.

Proof. Substituting (5.15) into (5.14b), we have for $\alpha \in [0,1]$ that

$$\begin{aligned} \|X(\alpha)S(\alpha)e - \mu(\alpha)e\| &\leq (1-\alpha)\theta\mu + \alpha^2\sigma\theta\mu \\ &\leq (1-\alpha+\sigma\alpha)\theta\mu \quad \text{since } \alpha \in [0,1] \\ &= \theta\mu(\alpha) \quad \text{from (5.7).} \end{aligned} \tag{5.16}$$

Hence the point $(x(\alpha), \lambda(\alpha), s(\alpha))$ satisfies the proximity condition for $\mathcal{N}_2(\theta)$.

We still have to check that $(x(\alpha), \lambda(\alpha), s(\alpha)) \in \mathcal{F}^o$. It is easy to verify that

$$Ax(\alpha) = b, \qquad A^T\lambda(\alpha) + s(\alpha) = c.$$

To check the positivity condition $(x(\alpha), s(\alpha)) > 0$, note first that $(x, s) = (x(0), s(0)) > 0$. It follows from (5.16) that

$$x_i(\alpha)s_i(\alpha) \geq (1-\theta)\mu(\alpha) = (1-\theta)(1-\alpha(1-\sigma))\mu > 0, \tag{5.17}$$

where the strict inequality is a consequence of $\theta \in (0,1)$, $\alpha \in (0,1]$, and $\sigma \in (0,1)$. Hence, we cannot have $x_i(\alpha) = 0$ or $s_i(\alpha) = 0$ for any index i when $\alpha \in [0,1]$. Therefore $(x(\alpha), s(\alpha)) > 0$ for all $\alpha \in [0,1]$, and so $(x(\alpha), \lambda(\alpha), s(\alpha)) \in \mathcal{F}^o$, as claimed. □

At this point, the proof of validity of Algorithm SPF is nearly complete. It remains only to check that our specific choice of parameters θ and σ, namely,

$$\theta = 0.4, \qquad \sigma = 1 - 0.4/\sqrt{n},$$

satisfies the condition (5.15) for all $n \geq 1$. We leave this as an exercise.

The Predictor-Corrector Method

In Algorithm SPF, we chose $\sigma_k \equiv \sigma$ to lie strictly between 0 and 1. This choice achieves the twin goals of improving centrality and reducing the duality measure μ into a single step. The predictor-corrector method,

Algorithm PC, takes two different kinds of steps to achieve each of these two goals. Successive iterations of Algorithm PC alternate between the two types of steps, which are

- *predictor* steps ($\sigma_k = 0$) to reduce μ,
- *corrector* steps ($\sigma_k = 1$) to improve centrality.

The other important ingredient in Algorithm PC is a *pair* of $\mathcal{N}_2$ neighborhoods, nested one inside the other. Even-index iterates ((x^k, λ^k, s^k), with k even) are confined to the inner neighborhood, whereas odd-index iterates are allowed to stray into the outer neighborhood, but not beyond.

The term *predictor-corrector* arose because of the analogy with predictor-corrector algorithms in ordinary differential equations (ODEs). These algorithms follow the solution trajectory of an initial-value ODE problem by alternating between predictor steps (which move along a tangent to the trajectory) and corrector steps (which move back toward the trajectory from the predicted point).

We examine the first two iterations of Algorithm PC, which suffice to illustrate the whole algorithm. Starting from a point (x^0, λ^0, s^0) in the inner neighborhood, we calculate a predictor step by setting $\sigma_0 = 0$. We move along this direction until we reach the boundary of the outer neighborhood. We stop at this point and define the new iterate (x^1, λ^1, s^1). A corrector step is now calculated by setting $\sigma_1 = 1$. A unit step along this direction ($\alpha = 1$) leads to a new iterate (x^2, λ^2, s^2) that is back inside the inner neighborhood. The two-step cycle then repeats, generating a sequence of iterates with the even-index iterates inside the inner neighborhood and the odd-index iterates on the boundary of the outer region. See Figure 5.2 for a depiction of this process.

Predictor steps reduce the value of μ by a factor of $(1 - \alpha)$, where α is the step length. Corrector steps leave μ unchanged, but by moving back into the inner neighborhood, they give the algorithm more room to maneuver on the next (predictor) iteration.

We obtain a formal specification of Algorithm PC by again filling in Framework PD of Chapter 1. For simplicity, we define the inner neighborhood to be $\mathcal{N}_2(0.25)$ and the outer neighborhood to be $\mathcal{N}_2(0.5)$. Other choices are possible, provided that the two neighborhoods are related to each other in a way that we define below.

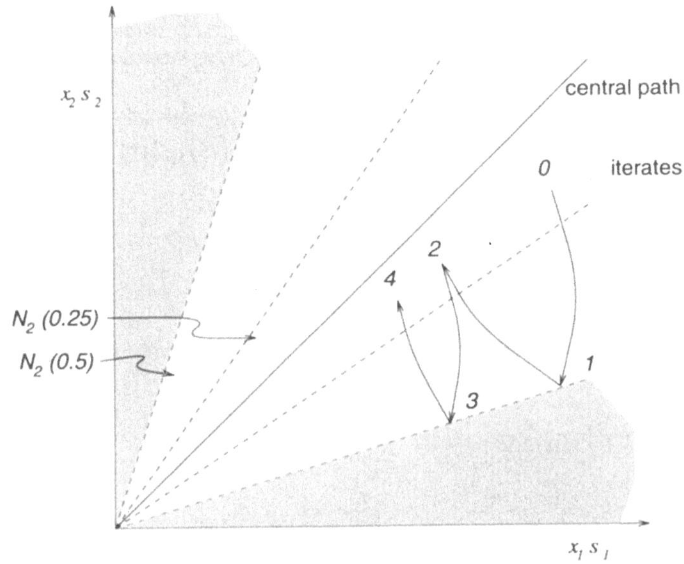

Figure 5.2. Iterates of Algorithm PC, plotted in (xs) space.

Algorithm PC
Given $(x^0, \lambda^0, s^0) \in \mathcal{N}_2(0.25)$;
for $k = 0, 1, 2, \ldots$
  if k is even
    (* predictor step *)
    solve (1.13) with $\sigma_k = 0$ to obtain $(\Delta x^k, \Delta\lambda^k, \Delta s^k)$;
    choose α_k as the largest value of α in $[0, 1]$ such that

$$(x^k(\alpha), \lambda^k(\alpha), s^k(\alpha)) \in \mathcal{N}_2(0.5); \tag{5.18}$$

    set $(x^{k+1}, \lambda^{k+1}, s^{k+1}) = (x^k(\alpha_k), \lambda^k(\alpha_k), s^k(\alpha_k))$;
  else
    (* corrector step *)
    solve (1.13) with $\sigma_k = 1$ to obtain $(\Delta x^k, \Delta\lambda^k, \Delta s^k)$;
    set $(x^{k+1}, \lambda^{k+1}, s^{k+1}) = (x^k, \lambda^k, s^k) + (\Delta x^k, \Delta\lambda^k, \Delta s^k)$;
  end (if)
end (for).

Our analysis of Algorithm PC is brief because most of the work has already been done in the analysis of Algorithm SPF. The behavior of each predictor step is described by the following lemma, which finds a lower bound on its step length and therefore an estimate of the reduction in μ.

Lemma 5.7 *Suppose that $(x, \lambda, s) \in \mathcal{N}_2(0.25)$, and let $(\Delta x, \Delta\lambda, \Delta s)$ be calculated from (1.12) with $\sigma = 0$. Then $(x(\alpha), \lambda(\alpha), s(\alpha)) \in \mathcal{N}_2(0.5)$ for all $\alpha \in [0, \bar{\alpha}]$, where*

$$\bar{\alpha} = \min\left(\tfrac{1}{2}, \left(\frac{\mu}{8\|\Delta X \Delta S e\|}\right)^{1/2}\right). \tag{5.19}$$

Hence, the predictor step has length at least $\bar{\alpha}$, and the new value of μ is at most $(1-\bar{\alpha})\mu$.

Proof. From (5.14a), we have

$$\begin{aligned}
&\|X(\alpha)S(\alpha)e - \mu(\alpha)e\| \\
&\quad\le (1-\alpha)\|XSe - \mu e\| + \alpha^2\|\Delta X \Delta S e\| \\
&\quad\le (1-\alpha)\|XSe - \mu e\| + \frac{\mu}{8\|\Delta X \Delta S e\|}\|\Delta X \Delta S e\| && \text{from (5.19)}, \\
&\quad\le \tfrac{1}{4}(1-\alpha)\mu + \frac{1}{8(1-\alpha)}(1-\alpha)\mu && \text{since } (x,\lambda,s) \in \mathcal{N}_2(0.25), \\
&\quad\le \tfrac{1}{4}(1-\alpha)\mu + \tfrac{1}{4}(1-\alpha)\mu && \text{since } \alpha \le \tfrac{1}{2}, \\
&\quad\le \tfrac{1}{2}\mu(\alpha) && \text{by (5.7), with } \sigma = 0.
\end{aligned}$$

Hence, the point $(x(\alpha), \lambda(\alpha), s(\alpha))$ satisfies the proximity condition for $\mathcal{N}_2(0.5)$. The remaining strict feasibility conditions can be verified as in the proof of Theorem 5.6. □

Lemma 5.4 can be used to find a lower bound on $\bar{\alpha}$. Setting $\theta = 0.25$ and $\sigma = 0$ in this result, we obtain

$$\frac{\mu}{8\|\Delta X \Delta S e\|} \ge \frac{2^{3/2}(1-0.25)}{8((0.25)^2 + n)} = \frac{3\sqrt{2}}{1+16n} \ge \frac{0.16}{n},$$

since $n \ge 1$. Hence, from (5.19) we have

$$\bar{\alpha} \ge \min\left(\tfrac{1}{2}, \left(\frac{0.16}{n}\right)^{1/2}\right) = \frac{0.4}{\sqrt{n}}.$$

Since predictor steps are taken at even-index iterates, this bound implies that

$$\mu_{k+1} \le \left(1 - \frac{0.4}{\sqrt{n}}\right)\mu_k, \qquad k = 0, 2, 4, \ldots. \tag{5.20}$$

Corrector steps are described by the following lemma, which shows that they return any point in $\mathcal{N}_2(0.5)$ to the inner neighborhood $\mathcal{N}_2(0.25)$ without changing the value of μ.

Lemma 5.8 *Suppose that* $(x, \lambda, s) \in \mathcal{N}_2(0.5)$, *and let* $(\Delta x, \Delta\lambda, \Delta s)$ *be calculated from* (1.12) *with* $\sigma = 1$. *Then we have*

$$(x(1), \lambda(1), s(1)) \in \mathcal{N}_2(0.25), \qquad \mu(1) = \mu.$$

Proof. By substituting $\sigma = 1$ into (5.7), we have immediately that $\mu(1) = \mu$. (In fact, $\mu(\alpha) = \mu$ for all $\alpha \in [0, 1]$.)

By substituting $\theta = 0.5$, $\alpha = 1$, and $\sigma = 1$ into (5.14b), we find that

$$\|X(1)S(1)e - \mu(1)e\| \le \mu/4 = \mu(1)/4.$$

Hence, $(x(1), \lambda(1), s(1))$ satisfies the proximity conditions for $\mathcal{N}_2(0.25)$. The proof is completed by verifying that $(x(1), \lambda(1), s(1))$ is also strictly feasible, which follows as in Theorem 5.6. □

As we see from this lemma, corrector iterations leave the value of the duality measure μ unchanged. However, because the predictor iterations achieve a substantial reduction in μ (5.20), we can prove the same kind of polynomial complexity result as for the short-step algorithm.

Theorem 5.9 *Given* $\epsilon > 0$, *suppose that the starting point* $(x^0, \lambda^0, s^0) \in \mathcal{N}_2(0.25)$ *in Algorithm PC has*

$$\mu_0 \le 1/\epsilon^\kappa$$

for some positive constant κ. *Then there is an index* K *with* $K = O(\sqrt{n} \log 1/\epsilon)$ *such that*

$$\mu_k \le \epsilon \qquad \textit{for all } k \ge K.$$

Proof. Combining (5.20) with Lemma 5.8, we have

$$\mu_{k+2} = \mu_{k+1} \le \left(1 - \frac{0.4}{\sqrt{n}}\right) \mu_k, \qquad k = 0, 2, 4, \ldots.$$

Hence, the reduction requirement (3.10) of Theorem 3.2 is *almost* satisfied when we set $\delta = 0.4$ and $\omega = 0.5$, except that the reduction in μ occurs over a span of *two* iterations instead of just one. The proof of Theorem 3.2 can be modified easily to handle this slightly different condition (as we showed in the exercises for Chapter 3) without affecting the conclusion of the theorem. □

The predictor-corrector algorithm is a definite improvement over the short-step algorithm because of the adaptivity that is built into the choice of predictor step length. In Algorithm SPF, the values of σ and α are fixed at conservative values so that they confine the iterates (x^k, λ^k, s^k) to the neighborhood $\mathcal{N}_2(\theta)$ under all circumstances. By contrast, the predictor step lengths of Algorithm PC are longer when the predictor direction is a good search direction, that is, when it produces a large reduction in μ without moving away from the central path too rapidly. During the final stages of the algorithm, the predictor directions become better and better, and Algorithm PC is eventually able to use step lengths close to 1. In fact, convergence of the duality measures μ_k to zero is superlinear, as we see in Chapter 7.

Despite its adaptivity, Algorithm PC is still restricted by the cramped nature of the $\mathcal{N}_2$ neighborhoods, particularly during early iterations when far from the solution. We now describe a long-step path-following algorithm that combines flexibility in the choice of step length with the use of a more liberal neighborhood $\mathcal{N}_{-\infty}(\gamma)$.

A Long-Step Path-Following Algorithm

Algorithm LPF generates a sequence of iterates in the neighborhood $\mathcal{N}_{-\infty}(\gamma)$, which, for small values of γ (say, $\gamma = 10^{-3}$), occupies most of the set $\mathcal{F}^o$ of strictly feasible points. At each iterate of Algorithm LPF, we choose the centering parameter σ_k to lie between the two fixed limits $\sigma_{\min}$ and $\sigma_{\max}$, where $0 < \sigma_{\min} < \sigma_{\max} < 1$. The search direction is, as usual, obtained by solving (1.13), and we choose the step length α_k to be as large as possible, subject to staying inside $\mathcal{N}_{-\infty}(\gamma)$.

A formal statement of the algorithm follows.

Algorithm LPF
Given γ, $\sigma_{\min}$, $\sigma_{\max}$ with $\gamma \in (0,1)$, $0 < \sigma_{\min} < \sigma_{\max} < 1$,
and $(x^0, \lambda^0, s^0) \in \mathcal{N}_{-\infty}(\gamma)$;
for $k = 0, 1, 2, \ldots$
choose $\sigma_k \in [\sigma_{\min}, \sigma_{\max}]$;
solve (1.13) to obtain $(\Delta x^k, \Delta \lambda^k, \Delta s^k)$;
choose α_k as the largest value of α in $[0,1]$ such that

$$(x^k(\alpha), \lambda^k(\alpha), s^k(\alpha)) \in \mathcal{N}_{-\infty}(\gamma); \tag{5.21}$$

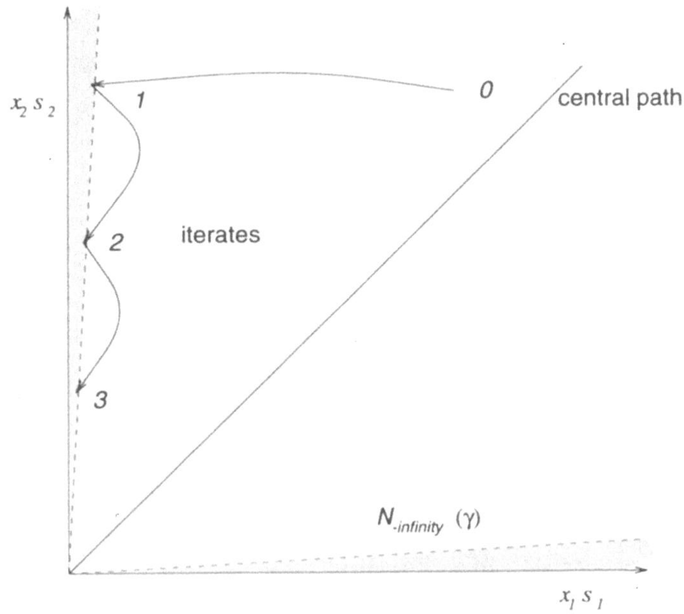

Figure 5.3. Iterates of Algorithm LPF, plotted in (xs) space.

set $(x^{k+1}, \lambda^{k+1}, s^{k+1}) = (x^k(\alpha_k), \lambda^k(\alpha_k), s^k(\alpha_k))$;

end (for).

Typical behavior of the algorithm is illustrated in Figure 5.3. As this figure shows (and the analysis confirms), the lower bound $\sigma_{\min}$ on the centering parameter ensures that each search direction starts out by moving off the boundary of $\mathcal{N}_{-\infty}(\gamma)$ and into the interior of this neighborhood. That is, small steps along the search direction improve the centrality. Larger values of α take us outside the neighborhood again, since the error of approximating the nonlinear system (5.1) by the linear step equations (1.12) becomes more pronounced as α increases. Still, we are guaranteed that a certain minimum step can be taken before we reach the boundary of $\mathcal{N}_{-\infty}(\gamma)$. Lemma 5.10 and Theorem 5.11 find a lower bound on α_k and a corresponding estimate of the reduction in μ at each iteration. Theorem 5.12 states the usual polynomial complexity result.

As an aside, we note that the representation of the neighborhoods $\mathcal{N}_2$ and $\mathcal{N}_{-\infty}$ in Figures 5.1, 5.2, and 5.3 is identical—both neighborhoods are demarcated by straight lines emanating from the origin. We see in the exercises that these two kinds of neighborhoods are closely related when

$n = 2$ but that this relationship breaks down for larger values of n. The two neighborhoods $\mathcal{N}_2$ and $\mathcal{N}_{-\infty}$ would have different shapes if we extended Figures 5.1, 5.2, or 5.3 into a third dimension.

Lemma 5.10 *If $(x, \lambda, s) \in \mathcal{N}_{-\infty}(\gamma)$, then*

$$\|\Delta X \Delta S e\| \leq 2^{-3/2}(1 + 1/\gamma) n\mu.$$

Proof. As in the proof of Lemma 5.4, we have

$$\|\Delta X \Delta S e\| \leq 2^{-3/2} \|(XS)^{-1/2}(-XSe + \sigma\mu e)\|^2.$$

Expanding the squared Euclidean norm and using such relationships as $x^T s = n\mu$ and $e^T e = n$, we obtain

$$\begin{aligned}
\|\Delta X \Delta S e\| &\leq 2^{-3/2} \left\| -(XS)^{1/2} e + \sigma\mu (XS)^{-1/2} e \right\|^2 \\
&\leq 2^{-3/2} \left[x^T s - 2\sigma\mu e^T e + \sigma^2\mu^2 \sum_{i=1}^{n} \frac{1}{x_i s_i} \right] \\
&\leq 2^{-3/2} \left[x^T s - 2\sigma\mu e^T e + \sigma^2\mu^2 \frac{n}{\gamma\mu} \right] \qquad \text{since } x_i s_i \geq \gamma\mu \\
&\leq 2^{-3/2} \left[1 - 2\sigma + \frac{\sigma^2}{\gamma} \right] n\mu \\
&\leq 2^{-3/2}(1 + 1/\gamma) n\mu,
\end{aligned}$$

as claimed. □

Theorem 5.11 *Given the parameters γ, $\sigma_{\min}$, and $\sigma_{\max}$ in Algorithm LPF, there is a constant δ independent of n such that*

$$\mu_{k+1} \leq \left(1 - \frac{\delta}{n}\right) \mu_k$$

for all $k \geq 0$.

Proof. We start by proving that

$$(x^k(\alpha), \lambda^k(\alpha), s^k(\alpha)) \in \mathcal{N}_{-\infty}(\gamma) \text{ for all } \alpha \in \left[0, 2^{3/2}\gamma \frac{1-\gamma}{1+\gamma} \frac{\sigma_k}{n}\right], \tag{5.22}$$

from which we deduce that α_k is bounded below as follows:

$$\alpha_k \geq 2^{3/2} \frac{\sigma_k}{n} \gamma \frac{1-\gamma}{1+\gamma}. \tag{5.23}$$

For any $i = 1, 2, \ldots, n$, we have from Lemma 5.10 that

$$|\Delta x_i^k \Delta s_i^k| \le \|\Delta X^k \Delta S^k e\|_2 \le 2^{-3/2}(1 + 1/\gamma)n\mu_k. \tag{5.24}$$

Using (5.8), we have from $x_i^k s_i^k \ge \gamma\mu_k$ and (5.24) that

$$\begin{aligned} x_i^k(\alpha)s_i^k(\alpha) &= (x_i^k + \alpha\Delta x_i^k)(s_i^k + \alpha\Delta s_i^k) \\ &= x_i^k s_i^k + \alpha(x_i^k \Delta s_i^k + s_i^k \Delta x_i^k) + \alpha^2 \Delta x_i^k \Delta s_i^k \\ &\ge x_i^k s_i^k(1 - \alpha) + \alpha\sigma_k\mu_k - \alpha^2|\Delta x_i^k \Delta s_i^k| \\ &\ge \gamma(1 - \alpha)\mu_k + \alpha\sigma_k\mu_k - \alpha^2 2^{-3/2}(1 + 1/\gamma)n\mu_k. \end{aligned}$$

Meanwhile, we have from (5.7) that

$$\mu_k(\alpha) = (1 - \alpha(1 - \sigma_k))\mu_k.$$

From these last two formulas, we can see that the proximity condition

$$x_i^k(\alpha)s_i^k(\alpha) \ge \gamma\mu_k(\alpha)$$

is satisfied, provided that

$$\gamma(1 - \alpha)\mu_k + \alpha\sigma_k\mu_k - \alpha^2 2^{-3/2}(1 + 1/\gamma)n\mu_k \ge \gamma(1 - \alpha + \alpha\sigma_k)\mu_k.$$

Rearranging this expression, we obtain

$$\alpha\sigma_k\mu_k(1 - \gamma) \ge \alpha^2 2^{-3/2} n\mu_k(1 + 1/\gamma),$$

which is true if

$$\alpha \le \frac{2^{3/2}}{n}\sigma_k\gamma\frac{1 - \gamma}{1 + \gamma}.$$

We have proved that $(x^k(\alpha), \lambda^k(\alpha), s^k(\alpha))$ satisfies the proximity condition for $\mathcal{N}_{-\infty}(\gamma)$ when α lies in the range stated in (5.22). We can also show, as in the proof of Theorem 5.6, that $(x^k(\alpha), \lambda^k(\alpha), s^k(\alpha)) \in \mathcal{F}^o$ for all α in the given range. Hence, we have proved (5.22) and therefore (5.23).

We complete the proof of the theorem by estimating the reduction in μ on the kth step. From (5.7) and (5.23), we have

$$\begin{aligned} \mu_{k+1} &= (1 - \alpha_k(1 - \sigma_k))\mu_k \\ &\le \left(1 - \frac{2^{3/2}}{n}\gamma\frac{1 - \gamma}{1 + \gamma}\sigma_k(1 - \sigma_k)\right)\mu_k. \end{aligned} \tag{5.25}$$

Now, the function $\sigma(1-\sigma)$ is a concave quadratic function of σ, so on any given interval it attains its minimum value at one of the endpoints. Hence, we have

$$\sigma_k(1-\sigma_k) \geq \min\left\{\sigma_{\min}(1-\sigma_{\min}), \sigma_{\max}(1-\sigma_{\max})\right\} \quad \text{for all } \sigma_k \in [\sigma_{\min}, \sigma_{\max}].$$

The proof is completed by substituting this estimate into (5.25) and setting

$$\delta = 2^{3/2}\gamma\frac{1-\gamma}{1+\gamma}\min\left\{\sigma_{\min}(1-\sigma_{\min}), \sigma_{\max}(1-\sigma_{\max})\right\}. \quad \square$$

The complexity result is an immediate consequence of Theorems 5.11 and 3.2.

Theorem 5.12 *Given $\epsilon > 0$ and $\gamma \in (0,1)$, suppose that the starting point $(x^0, \lambda^0, s^0) \in \mathcal{N}_{-\infty}(\gamma)$ in Algorithm LPF has*

$$\mu_0 \leq 1/\epsilon^\kappa$$

for some positive constant κ. Then there is an index K with $K = O(n \log 1/\epsilon)$ such that

$$\mu_k \leq \epsilon \qquad \text{for all } k \geq K.$$

Limit Points of the Iteration Sequence

The convergence results of this chapter have focused thus far on convergence of the sequence $\{\mu_k\}$ to zero, without saying anything about the sequence of iterates $\{(x^k, \lambda^k, s^k)\}$. The behavior of the iterate sequence is a little more complicated than one might expect. The main issue is to show that the sequence $\{(x^k, s^k)\}$ has a limit point, because we can construct a primal-dual solution from any such limit point by the following argument: If $\mathcal{K}$ is the subsequence for which $\lim_{k\in\mathcal{K}}(x^k, s^k) = (x^*, s^*)$, we have for all $k \in \mathcal{K}$ that

$$Ax^k = b, \qquad c - s^k \in \text{Range}(A^T), \qquad (x^k, s^k) > 0.$$

Taking limits and using the facts that $\text{Range}(A^T)$ is closed and $\mu_k \downarrow 0$, we find that (x^*, s^*) satisfies

$$Ax^* = b, \qquad c - s^* \in \text{Range}(A^T), \qquad (x^*, s^*) \geq 0, \qquad (x^*)^T s^* = 0.$$

Hence, $c - s^* = A\lambda^*$ for some λ^*. Comparing these conditions with the KKT conditions (1.4), we conclude that $(x^*, \lambda^*, s^*) \in \Omega$ as claimed.

In this section, we look at the limiting behavior of the sequence $\{(x^k, s^k)\}$ generated by each algorithm in this chapter. We show that the sequence is bounded and therefore has at least one limit point. Further, all limit points correspond to strictly complementary solutions, that is, solutions (x^*, λ^*, s^*) for which

$$x_i^* > 0 \ \ (i \in \mathcal{B}), \qquad s_i^* > 0 \ \ (i \in \mathcal{N}), \tag{5.26}$$

where $\mathcal{B} \cup \mathcal{N}$ is the partition of $\{1, 2, \ldots, n\}$ defined in (2.11).

Lemma 5.13 *Let $\mu_0 > 0$ and $\gamma \in (0, 1)$. Then for all points (x, λ, s) with*

$$(x, \lambda, s) \in \mathcal{N}_{-\infty}(\gamma) \subset \mathcal{F}^o, \qquad \mu \le \mu_0 \tag{5.27}$$

(where $\mu = x^T s/n$), there are constants C_0 and C_3 such that

$$\|(x, s)\| \le C_0, \tag{5.28}$$

$$0 < x_i \le \mu/C_3 \ \ (i \in \mathcal{N}), \qquad 0 < s_i \le \mu/C_3 \ \ (i \in \mathcal{B}), \tag{5.29}$$

$$s_i \ge C_3\gamma \ \ (i \in \mathcal{N}), \qquad x_i \ge C_3\gamma \ \ (i \in \mathcal{B}). \tag{5.30}$$

Proof. The first result (5.28) follows immediately from Lemma 2.5 if we set $K = n\mu_0$.

For (5.29) and (5.30), let (x^*, λ^*, s^*) be any primal-dual solution. Since this solution and the point (x, λ, s) are both feasible, we have

$$Ax = Ax^* = b, \qquad A^T\lambda + s = A^T\lambda^* + s^* = c,$$

and therefore

$$(x - x^*)^T(s - s^*) = -(x - x^*)^T A^T(\lambda - \lambda^*) = 0.$$

Since (2.11) implies that $x_i^* = 0$ for $i \in \mathcal{N}$ and $s_i^* = 0$ for $i \in \mathcal{B}$, we can rearrange this expression to obtain

$$n\mu = x^T s^* + s^T x^* = \sum_{i \in \mathcal{N}} x_i s_i^* + \sum_{i \in \mathcal{B}} s_i x_i^*.$$

Since each term in the summations is nonnegative, each term is bounded by $n\mu$. Hence, for any $i \in \mathcal{N}$ with $s_i^* > 0$, we have

$$0 < x_i s_i^* \le n\mu \ \Rightarrow \ 0 < x_i \le \frac{n}{s_i^*}\mu. \tag{5.31}$$

Since the expression (5.31) holds for *any* solution (x^*, λ^*, s^*) with $s_i^* > 0$, we choose the one that yields the tightest bound on x_i, that is,

$$0 < x_i \leq \frac{n}{\sup_{(\lambda^*,s^*)\in\Omega_D} s_i^*}\,\mu.$$

Taking the maximum of this bound over the indices $i \in \mathcal{N}$, we obtain

$$\max_{i\in\mathcal{N}} x_i \leq \frac{n}{\min_{i\in\mathcal{N}} \sup_{(\lambda^*,s^*)\in\Omega_D} s_i^*}\,\mu.$$

Similarly,

$$0 < \max_{i\in\mathcal{B}} s_i \leq \frac{n}{\min_{i\in\mathcal{B}} \sup_{x^*\in\Omega_P} x_i^*}\,\mu.$$

Combining the two estimates, we obtain

$$\begin{aligned}\max\left(\max_{i\in\mathcal{N}} x_i, \max_{i\in\mathcal{B}} s_i\right) &\leq n\left[\min\left(\min_{i\in\mathcal{B}}\sup_{x^*\in\Omega_P} x_i^*, \min_{i\in\mathcal{N}}\sup_{(\lambda^*,s^*)\in\Omega_D} s_i^*\right)\right]^{-1}\mu \\ &= \frac{n}{\epsilon(A,b,c)}\,\mu,\end{aligned}$$

where $\epsilon(A,b,c)$ was defined in (3.5). The result (5.29) follows immediately when we set

$$C_3 = \frac{\epsilon(A,b,c)}{n}. \tag{5.32}$$

Existence of a strictly complementary solution (Theorem 2.4) guarantees that $\epsilon(A,b,c) > 0$, so C_3 is positive.

Finally, since $(x,\lambda,s) \in \mathcal{N}_{-\infty}(\gamma)$, we have $x_i s_i \geq \gamma\mu$ for all $i = 1, 2, \ldots, n$. Hence, we have from (5.29) that

$$s_i \geq \frac{\gamma\mu}{x_i} \geq \frac{\gamma\mu}{\mu/C_3} = C_3\gamma \qquad \text{for all } i \in \mathcal{N}.$$

In the same way, we can show that $x_i \geq C_3\gamma$ for $i \in \mathcal{B}$, proving (5.30). □

Theorem 5.14 *Let $\{(x^k, \lambda^k, s^k)\}$ be a sequence of iterates generated by Algorithm SPF, PC, or LPF, and suppose that $\mu_k \downarrow 0$ as $k \to \infty$. Then the sequence $\{(x^k, s^k)\}$ is bounded and therefore has at least one limit point. Each limit point corresponds to a strictly complementary primal-dual solution.*

Proof. All three algorithms confine their iteration sequences to a neighborhood $\mathcal{N}_{-\infty}(\gamma)$ for some $\gamma > 0$. In Algorithm SPF, we have

$$(x^k, \lambda^k, s^k) \in \mathcal{N}_2(0.4) \subset \mathcal{N}_{-\infty}(0.4).$$

In Algorithm PC, all iterates belong to $\mathcal{N}_2(0.5)$, which is a subset of $\mathcal{N}_{-\infty}(0.5)$, whereas in Algorithm LPF, the value of γ is chosen explicitly. Also, the sequences $\{\mu_k\}$ are nonincreasing for each method; in particular, $\mu_k \leq \mu_0$ for all $k \geq 0$. Hence each iterate (x^k, λ^k, s^k) satisfies the assumptions of Lemma 5.13.

Boundedness of $\{(x^k, s^k)\}$ follows from (5.28). If (x^*, s^*) is a limit point, we can find $\lambda^* \in \mathbb{R}^m$ such that $(x^*, \lambda^*, s^*) \in \Omega$ (see the discussion above). Because of (5.30), we must have

$$s_i^* \geq C_3\gamma > 0 \ \ (i \in \mathcal{N}), \qquad x_i^* \geq C_3\gamma > 0 \ \ (i \in \mathcal{B}),$$

so the solution is strictly complementary. □

When the problem has a unique primal-dual solution (x^*, λ^*, s^*), it follows immediately from Theorem 5.14 that the iteration sequences for all three algorithms converge to this point.

Proof of Lemma 5.3

We return to the technical lemma stated earlier in the chapter, which claimed that for any vector pair u, v with $u^T v \geq 0$, we have

$$\|UVe\| \leq 2^{-3/2}\|u + v\|^2.$$

First, note that for any two scalars α and β with $\alpha\beta \geq 0$, we have from the algebraic-geometric mean inequality that

$$\sqrt{|\alpha\beta|} \leq \frac{1}{2}|\alpha + \beta|. \tag{5.33}$$

Since $u^T v \geq 0$, we have

$$0 \leq u^T v = \sum_{u_i v_i \geq 0} u_i v_i + \sum_{u_i v_i < 0} u_i v_i = \sum_{i \in \mathcal{P}} |u_i v_i| - \sum_{i \in \mathcal{M}} |u_i v_i|, \tag{5.34}$$

where we partitioned the index set $\{1, 2, \ldots, n\}$ as

$$\mathcal{P} = \{i \,|\, u_i v_i \geq 0\}, \qquad \mathcal{M} = \{i \,|\, u_i v_i < 0\}.$$

Now,

$$
\begin{aligned}
\|UVe\| &= \left(\|[u_i v_i]_{i\in\mathcal{P}}\|^2 + \|[u_i v_i]_{i\in\mathcal{M}}\|^2 \right)^{1/2} \\
&\le \left(\|[u_i v_i]_{i\in\mathcal{P}}\|_1^2 + \|[u_i v_i]_{i\in\mathcal{M}}\|_1^2 \right)^{1/2} && \text{since } \|\cdot\|_2 \le \|\cdot\|_1 \\
&\le \left(2\,\|[u_i v_i]_{i\in\mathcal{P}}\|_1^2 \right)^{1/2} && \text{from (5.34)} \\
&\le \sqrt{2} \left\| \left[\frac{1}{4}(u_i+v_i)^2 \right]_{i\in\mathcal{P}} \right\|_1 && \text{from (5.33)} \\
&= 2^{-3/2} \sum_{i\in\mathcal{P}} (u_i+v_i)^2 \\
&\le 2^{-3/2} \sum_{i=1}^{n} (u_i+v_i)^2 \\
&= 2^{-3/2} \|u+v\|^2,
\end{aligned}
$$

completing the proof.

Notes and References

Xu [152] has described a way to avoid the tight confines of the neighborhood $\mathcal{N}_2(\theta)$, $\theta < 1$ without sacrificing $O(\sqrt{n}L)$ complexity. He uses the neighborhood $\mathcal{N}_2(\theta) \cap \mathcal{N}_{-\infty}(\gamma)$, where $\gamma > 0$ is small but θ may be *much larger* than 1. Steps in this neighborhood can be almost as long as in the neighborhood $\mathcal{N}_{-\infty}$ alone, and the strategy has been successfully implemented by Xu, Hung, and Ye [153].

Lemma 5.13 and Theorem 5.14 are due to Güler and Ye [50]. The set of limit points of $\{(x^k, s^k)\}$ actually forms a continuum (see Tapia, Zhang, and Ye [126, Theorem 4.1]). Convergence of $\{(x^k, s^k)\}$ is discussed further in Chapters 6 and 7.

Exercises

1. Check that the choices of θ and σ used in Algorithm SPF satisfy the relationship (5.15).

2. Prove (5.6). Does this result still hold for the infeasible step (1.20)?

3. For a given $\theta \in (0, 1)$, the problem of finding a value of σ that satisfies (5.15) while maximizing the decrease in μ for a unit step can be posed as a simple constrained optimization problem. Write down this problem. Does this problem have a solution for all $\theta \in (0, 1)$? Explain.

4. Theorem 5.6 shows that the unit step for Algorithm SPF keeps the new iterate inside the neighborhood $\mathcal{N}_2(\theta)$. Express the problem of finding the maximal α such that $\|X(\alpha)S(\alpha)e - \mu(\alpha)e\|_2 \le \theta\mu(\alpha)$ as a quartic polynomial in α.

5. In Algorithms SPF and LPF, is there any benefit to taking a step of length *longer* than 1, if such a step remains inside the required neighborhood?

6. In Algorithm PC, we chose the inner and outer neighborhoods to be $\mathcal{N}_2(0.25)$ and $\mathcal{N}_2(0.5)$, respectively. In this exercise, you are asked to consider more general neighborhoods $\mathcal{N}_2(\theta_{\text{in}})$ and $\mathcal{N}_2(\theta_{\text{out}})$ and look for conditions on the scalars θ_{in} and θ_{out} for which the analysis of the algorithm continues to hold.

   (i) Redefine the lower bound $\bar{\alpha}$ on the predictor step length obtained in Lemma 5.7 for arbitrary values of θ_{in} and θ_{out}. That is, find a value η such that
   $$\bar{\alpha} = \min\left(\frac{1}{2}, \left(\frac{\eta\mu}{\|\Delta X \Delta S e\|}\right)^{1/2}\right).$$

   (ii) Modify the analysis of Lemma 5.8 to find the relationship between θ_{in} and θ_{out} that guarantees that the corrector step returns to the inner neighborhood.

   (iii) What is the largest value of θ_{out} for which the θ_{in} of part (ii) satisfies $\theta_{\text{in}} \in (0, \theta_{\text{out}})$?

7. Prove that when $n = 2$, the neighborhoods $\mathcal{N}_2(\theta)$ and $\mathcal{N}_{-\infty}(1-\theta/\sqrt{2})$ are identical. Does a similar relationship hold when $n = 3$?

Chapter 6

Infeasible-Interior-Point Algorithms

In Chapter 5, we looked at three path-following algorithms that start from a strictly feasible point (x^0, λ^0, s^0) in some neighborhood of the central path. Often, it is not easy to find a starting point that satisfies these conditions. Indeed, there are perfectly valid linear programs for which such points do not even exist. Consider, for instance, the problem

$$\min\ 2x_1 + x_2 \qquad \text{subject to } x_1 + x_2 + x_3 = 5,\ x_1 + x_3 = 5,\ x \geq 0,$$

for which the primal feasible set is $\{(\beta, 0, 5-\beta) \mid \beta \in [0, 5]\}$. Because $x_2 = 0$ for primal feasible points, the set $\mathcal{F}^o$ of strictly feasible points is empty. More generally, linear programs obtained by transformation of general problems to standard form often have no strictly feasible points, as we see in the exercises.

One way to avoid this difficulty is to embed the given linear program in a slightly larger problem for which a strictly feasible point is easy to identify. The homogeneous self-dual reformulation is a particularly useful embedding tool; we discuss it further in Chapter 9.

In this chapter, we describe a different approach: an algorithm that does not require the initial point to be strictly feasible but requires only that its x and s components be strictly positive. Algorithm IPF, specified later, is an infeasible-path-following variant of Algorithm LPF from Chapter 5. This algorithm is particularly interesting because it is not far removed from the practical algorithms that are used as the basis of primal-dual software. In general, all iterates (x^k, λ^k, s^k) generated by Algorithm IPF are infeasible, although its limit points are, of course, feasible (and optimal). The method works even when the set $\mathcal{F}^o$ of strictly feasible points is empty.

The steps in Algorithm IPF are obtained by solving the modified Newton equations (1.20) for values of the centering parameter σ bounded away from zero. The neighborhood to which the iterates are confined is an extension of $\mathcal{N}_{-\infty}(\gamma)$ from Chapter 5 that admits infeasible points. Two new conditions are imposed on the choice of step length α. First, the amount by which the constraints $A^T\lambda + s = c$ and $Ax = b$ are violated is required to decrease at least as rapidly as the duality measure μ. Second, we enforce an Armijo-like condition to ensure that μ decreases by at least some small fraction of the predicted decrease at every step.

In the analysis below, we prove that the sequence of duality measures μ_k produced by Algorithm IPF converges monotonically to zero from any starting point for which $(x^0, s^0) > 0$. If the starting point is chosen to have the special form $(x^0, \lambda^0, s^0) = (\zeta e, 0, \zeta e)$ for ζ sufficiently large, we obtain a polynomial complexity result: a point for which $\mu_k \leq \epsilon$ is obtained in $O(n^2|\log \epsilon|)$ iterations. At the end of the chapter, we prove some results on convergence of the iteration sequence, which are similar to those obtained in Chapter 5.

The statement of Algorithm IPF is not much more complicated than Algorithm LPF, but the analysis is more technical, although it takes up less space than perhaps might be expected. We include all the details, to show that a complete, self-contained analysis of an infeasible-interior-point algorithm need not be mysterious or especially complicated.

The impetus for infeasible-interior-point methods came, naturally enough, from researchers who were trying to develop practical interior-point implementations. The pre-1989 literature assumed a feasible initial point, although some papers did show how a simple reformulation of the problem could turn an infeasible starting point into a feasible starting point; see Megiddo [85], Kojima, Mizuno, and Yoshise [67], and Monteiro and Adler [94]. This reformulation, in which a few extra rows and columns were included in the constraint matrix to soak up the infeasibility in the given point, created some difficulties for the implementations, chiefly in the choice of weighting terms for the new primal and dual variables. The required size of these weights was difficult to determine a priori and the computational performance was, unfortunately, sensitive to their values (see McShane, Monma, and Shanno [84]). Lustig [76] examined the limiting directions generated by the standard primal-dual equations when these weights were driven to ∞, and proposed an algorithm based on these limiting directions in which the weights did not appear at all. Subsequently, Lustig, Marsten, and Shanno [77] showed that Lustig's limiting directions coincided with the directions obtained from the infeasible step equations (1.20) (see also (6.4) below).

Although codes based on the infeasible step equations were highly successful from the start, some researchers doubted that a global and polynomial convergence analysis was possible for an infeasible-interior-point algorithm. Breakthroughs were reported by Kojima, Megiddo, and Mizuno [64], who proved global convergence, and by Zhang [160], who proved polynomial complexity of a long-step path-following infeasible method. Other familiar properties of feasible interior-point methods were soon extended to the infeasible-interior-point framework, including the extension to LCPs (Zhang [160]) and superlinearly convergent variants (Potra [108], Wright [147]).

The Algorithm

Let us start by recalling the definitions of the residuals from Chapter 1:

$$r_b = Ax - b, \qquad r_c = A^T\lambda + s - c. \tag{6.1}$$

When these vectors are calculated at the point $(x, \lambda, s) = (x^k, \lambda^k, s^k)$, we denote them by r_b^k and r_c^k. As mentioned above, the algorithm described in this chapter works with a central path neighborhood $\mathcal{N}_{-\infty}(\gamma, \beta)$, an extension of $\mathcal{N}_{-\infty}(\gamma)$ from (5.4) that includes infeasible points. This neighborhood is defined by

$$\begin{aligned}\mathcal{N}_{-\infty}(\gamma, \beta) = \{(x, \lambda, s) \mid \|(r_b, r_c)\| \le [\|(r_b^0, r_c^0)\|/\mu_0]\beta\mu, \\ (x, s) > 0, \quad x_i s_i \ge \gamma\mu,\ i = 1, 2, \dots, n\},\end{aligned} \tag{6.2}$$

where $\gamma \in (0, 1)$ and $\beta \ge 1$ are given parameters and (r_b^0, r_c^0) and μ_0 are evaluated at the starting point (x^0, λ^0, s^0). Note that we must have

$$\beta \ge 1 \tag{6.3}$$

to ensure that the initial point (x^0, λ^0, s^0) belongs to $\mathcal{N}_{-\infty}(\gamma, \beta)$.

For all points in the neighborhood $\mathcal{N}_{-\infty}(\gamma, \beta)$, the infeasibility is uniformly bounded by some multiple of the duality measure μ. By forcing μ_k to zero monotonically and restricting all iterates (x^k, λ^k, s^k) to the neighborhood $\mathcal{N}_{-\infty}(\gamma, \beta)$, Algorithm IPF ensures that $r_b^k \to 0$ and $r_c^k \to 0$ as $k \to \infty$. The other condition in (6.2)—$x_i s_i \ge \gamma\mu$—serves the same purpose as in the feasible neighborhood $\mathcal{N}_{-\infty}(\gamma)$. It keeps the pairwise products $x_i s_i$ roughly in balance and prevents the Newton-like search directions from being distorted by components of (x, s) that approach zero prematurely.

Algorithm IPF
Given γ, β, $\sigma_{\min}$, $\sigma_{\max}$ with $\gamma \in (0,1)$, $\beta \geq 1$, and $0 < \sigma_{\min} < \sigma_{\max} \leq 0.5$;
choose (x^0, λ^0, s^0) with $(x^0, s^0) > 0$;
for $k = 0, 1, 2, \ldots$
choose $\sigma_k \in [\sigma_{\min}, \sigma_{\max}]$ and solve

$$\begin{bmatrix} 0 & A^T & I \\ A & 0 & 0 \\ S^k & 0 & X^k \end{bmatrix} \begin{bmatrix} \Delta x^k \\ \Delta \lambda^k \\ \Delta s^k \end{bmatrix} = \begin{bmatrix} -r_c^k \\ -r_b^k \\ -X^k S^k e + \sigma_k \mu_k e \end{bmatrix}; \tag{6.4}$$

choose α_k as the largest value of α in $[0,1]$ such that

$$(x^k(\alpha), \lambda^k(\alpha), s^k(\alpha)) \in \mathcal{N}_{-\infty}(\gamma, \beta) \tag{6.5}$$

and the following Armijo condition holds:

$$\mu_k(\alpha) \leq (1 - .01\alpha)\mu_k; \tag{6.6}$$

set $(x^{k+1}, \lambda^{k+1}, s^{k+1}) = (x^k(\alpha_k), \lambda^k(\alpha_k), s^k(\alpha_k))$;
end (for).

Following (5.5), we have used the notation

$$\begin{aligned} (x(\alpha), \lambda(\alpha), s(\alpha)) &= (x, \lambda, s) + \alpha(\Delta x, \Delta \lambda, \Delta s), \\ \mu(\alpha) &= x(\alpha)^T s(\alpha)/n. \end{aligned}$$

We can allow some flexibility in the choice of α_k. Instead of choosing it to be the *largest* value for which the conditions (6.5) and (6.6) are satisfied (although this value is easy to find), we could allow smaller values that still satisfy the two conditions, provided they are not too much smaller than the largest value. Smaller values might actually be better; for instance, we could choose α_k to *minimize* $\mu(\alpha)$ rather than just satisfy the Armijo condition (6.6). We revisit this point in the exercises and restrict the present discussion to the largest value for the sake of simplicity.

Before proceeding, we introduce the useful scalar quantity ν_k defined by

$$\nu_k = \prod_{j=0}^{k-1} (1 - \alpha_j) \qquad (\nu_0 = 1). \tag{6.7}$$

Because the first two components of the function F in (1.5) are linear, we have

$$\begin{aligned}(r_b^k, r_c^k) &= (1-\alpha_{k-1})(r_b^{k-1}, r_c^{k-1}) \\ &= (1-\alpha_{k-1})\cdots(1-\alpha_0)(r_b^0, r_c^0) \\ &= \nu_k(r_b^0, r_c^0), \end{aligned} \tag{6.8}$$

while, since (x^k, λ^k, s^k) belongs to $\mathcal{N}_{-\infty}(\gamma, \beta)$, we have from (6.8) that

$$\nu_k \|(r_b^0, r_c^0)\|/\mu_k = \|(r_b^k, r_c^k)\|/\mu_k \le \beta \|(r_b^0, r_c^0)\|/\mu_0.$$

Provided that $(r_b^0, r_c^0) \neq 0$, it follows from this inequality that

$$\nu_k \le \beta \mu_k / \mu_0. \tag{6.9}$$

In the feasible case of $(r_b^0, r_c^0) = 0$, all iterates (x^k, λ^k, s^k) are strictly feasible. Algorithm IPF reduces in this case to a variant of Algorithm LPF, the long-step path-following algorithm from Chapter 5. For simplicity, we ignore this special case for the rest of the chapter and assume that the initial point (x^0, λ^0, s^0) is infeasible.

Convergence of Algorithm IPF

Convergence of this algorithm is proved by showing that there is a constant $\bar{\alpha} > 0$ such that $\alpha_k \ge \bar{\alpha}$ for every k. Because of the condition (6.6), it follows that

$$\mu_{k+1} \le (1 - .01\alpha_k)\mu_k \le (1 - .01\bar{\alpha})\mu_k \qquad \text{for all } k \ge 0, \tag{6.10}$$

so that the sequence $\{\mu_k\}$ of duality gaps converges Q-linearly to zero. Because of (6.8), we also have

$$\|(r_b^{k+1}, r_c^{k+1})\| \le (1 - \bar{\alpha}) \|(r_b^k, r_c^k)\|, \tag{6.11}$$

so the sequence of residual norms also converges to zero.

This discussion is formalized in the following global convergence theorem, which we prove later in the chapter.

Theorem 6.1 *The sequence $\{\mu_k\}$ generated by Algorithm IPF converges Q-linearly to zero, and the sequence of residual norms $\{\|(r_b^k, r_c^k)\|\}$ converges R-linearly to zero.*

Polynomial complexity of the algorithm follows from the fact that the lower bound $\bar{\alpha}$ on the step length is an inverse polynomial function of n if we choose the starting point to be

$$(x^0, \lambda^0, s^0) = (\zeta e, 0, \zeta e), \tag{6.12}$$

where ζ is a scalar for which

$$\|(x^*, s^*)\|_\infty \leq \zeta \tag{6.13}$$

for some primal-dual solution (x^*, λ^*, s^*). Of course, we usually don't know $\|(x^*, s^*)\|_\infty$ because we don't know the optimal solutions a priori. However, these conditions are still relevant to computational practice for the following reason: Computationally speaking, well-centered starting points for which the ratio

$$\|(r_b^0, r_c^0)\|/\mu_0 \tag{6.14}$$

is small lead to faster convergence than do poorly centered points that are much closer to the solution set. The point (6.12) satisfies these criteria. It is perfectly centered, and the ratios (6.14) vary like $1/\zeta$.

The precise statement of the complexity result is as follows. Again, the proof appears later in the chapter.

Theorem 6.2 *Let $\epsilon > 0$ be given. Suppose that we use the starting point $(x^0, \lambda^0, s^0) = (\zeta e, 0, \zeta e)$, where ζ is defined in Lemma 6.4. Suppose that the value of ζ for our particular linear programming instance satisfies*

$$\zeta^2 \leq \frac{C}{\epsilon^\kappa}$$

for some positive constants C and κ. Then there is an index K with

$$K = O\left(n^2 |\log \epsilon|\right)$$

such that the iterates $\{(x^k, \lambda^k, s^k)\}$ generated by Algorithm IPF satisfy

$$\mu_k \leq \epsilon \qquad \text{for all } k \geq K.$$

We lay out the technical basis for the proofs of Theorems 6.1 and 6.2 in the following three sections. The analysis takes the form of five lemmas (Lemmas 6.3–6.7) whose proofs require a lot of manipulation of inequalities but no conceptually deep mathematics. The treatment below draws on a number of sources, including Kojima [62], Zhang [160], and Wright [146].

Lemmas 6.3 and 6.5 combined with the estimate of α_k in Lemma 6.7 give the proof of the global convergence result, Theorem 6.1. A similar path is followed to the complexity result, Theorem 6.2. Lemma 6.7 is combined with Lemmas 6.4 and 6.6, which are minor variants of Lemmas 6.3 and 6.5, respectively.

Technical Results I: Bounds on $\nu_k\|(x^k, s^k)\|$

The following observation is useful in several places in the analysis that follows. If $(\bar{x}, \bar{\lambda}, \bar{s})$ is any vector that satisfies the conditions

$$A\bar{x} = 0, \qquad A^T\bar{\lambda} + \bar{s} = 0, \tag{6.15}$$

then

$$\bar{x}^T\bar{s} = -\bar{x}^T A^T \bar{\lambda} = 0. \tag{6.16}$$

The first result finds a bound on $\nu_k\|(x^k, s^k)\|$. In the proof of this result and all other results in this chapter, we omit the iteration index k on vector quantities for reasons of clarity.

Lemma 6.3 *There is a positive constant C_1 such that*

$$\nu_k\|(x^k, s^k)\|_1 \le C_1\mu_k \qquad \textit{for all } k \ge 0. \tag{6.17}$$

Proof. Let (x^*, λ^*, s^*) be any primal-dual solution and $(\bar{x}, \bar{\lambda}, \bar{s})$ be the vector defined by

$$(\bar{x}, \bar{\lambda}, \bar{s}) = \nu_k(x^0, \lambda^0, s^0) + (1-\nu_k)(x^*, \lambda^*, s^*) - (x, \lambda, s). \tag{6.18}$$

It is not hard to check that $(\bar{x}, \bar{\lambda}, \bar{s})$ satisfies the conditions (6.15). Hence, from (6.16), we have

$$\begin{aligned} 0 &= \bar{x}^T\bar{s} \\ &= \left(\nu_k x^0 + (1-\nu_k)x^* - x\right)^T \left(\nu_k s^0 + (1-\nu_k)s^* - s\right) \\ &= \nu_k^2 (x^0)^T s^0 + (1-\nu_k)^2 (x^*)^T s^* + \nu_k(1-\nu_k)\left((x^0)^T s^* + (s^0)^T x^*\right) \\ &\quad + x^T s - \nu_k(s^T x^0 + x^T s^0) - (1-\nu_k)(s^T x^* + x^T s^*). \end{aligned} \tag{6.19}$$

Because all the components of (x, s) and (x^*, s^*) are nonnegative, we have $s^T x^* + x^T s^* \ge 0$. Moreover, since (x^*, λ^*, s^*) is a solution, we have $(x^*)^T s^* = 0$. By using these observations in (6.19) and rearranging, we obtain

$$\begin{aligned} &\nu_k(s^T x^0 + x^T s^0) \\ &\le \nu_k^2 (x^0)^T s^0 + x^T s + \nu_k(1-\nu_k)\left((x^0)^T s^* + (s^0)^T x^*\right). \end{aligned} \tag{6.20}$$

Let us now define the constant ξ by

$$\xi = \min_{i=1,2,\ldots,n} \min(x_i^0, s_i^0). \tag{6.21}$$

Because $(x^0, s^0) > 0$, we have $\xi > 0$. Moreover, since $(x,s) > 0$, it is easy to verify that

$$\xi \|(x,s)\|_1 \le (\min_i s_i^0)\|x\|_1 + (\min_i x_i^0)\|s\|_1 \le x^T s^0 + s^T x^0.$$

Now from (6.20), and using $\nu_k \in (0,1)$ and the definition $\mu_k = x^T s/n$, we have

$$\begin{aligned} \xi \nu_k \|(x,s)\|_1 &\le \nu_k^2 n\mu_0 + n\mu_k + \nu_k(1-\nu_k)\left(\|x^0\|_\infty \|s^*\|_1 + \|s^0\|_\infty \|x^*\|_1\right) \\ &\le \nu_k n\mu_0 + n\mu_k + \nu_k \|(x^0,s^0)\|_\infty \|(x^*,s^*)\|_1. \end{aligned} \tag{6.22}$$

We can now substitute for ν_k from (6.9) to obtain

$$\xi \nu_k \|(x,s)\|_1 \le \beta n\mu_k + n\mu_k + \beta\mu_k \|(x^0,s^0)\|_\infty \|(x^*,s^*)\|_1/\mu_0.$$

The result is obtained by setting

$$C_1 = \xi^{-1}\left[\beta n + n + \beta \|(x^0,s^0)\|_\infty \|(x^*,s^*)\|_1/\mu_0\right]. \quad \square$$

The next result is a special case of Lemma 6.3 in which we choose the starting point to satisfy (6.12) and (6.13). We determine the explicit dependence of the upper bound (6.17) on n. (Polynomial complexity results always require bounds in which the dependence on n is explicit.)

Lemma 6.4 *Suppose that the initial point is chosen to satisfy* (6.12) *and* (6.13). *Then for any iterate* (x^k, λ^k, s^k), *we have*

$$\zeta \nu_k \|(x^k, s^k)\|_1 \le 4\beta n\mu_k. \tag{6.23}$$

Proof. From the definition (6.21), we have $\xi = \zeta$. By our choice of ζ and (x^0, λ^0, s^0), we also have

$$\begin{aligned} \|(x^0,s^0)\|_\infty &= \zeta, \\ \|(x^*,s^*)\|_1 \le 2n\|(x^*,s^*)\|_\infty &\le 2n\zeta, \\ \mu_0 = (x^0)^T s^0/n &= \zeta^2. \end{aligned}$$

Substituting these values into (6.22) and using (6.9) and (6.3), we obtain

$$\begin{aligned}\zeta\nu_k\|(x,s)\|_1 &\le n\nu_k\mu_0 + n\mu_k + \nu_k(2n\zeta^2)\\ &\le \beta n\mu_k + n\mu_k + \beta(\mu_k/\mu_0)(2n\zeta^2)\\ &\le 4\beta n\mu_k,\end{aligned}$$

as required. □

Technical Results II: Bounds on $(D^k)^{-1}\Delta x^k$ and $D^k\Delta s^k$

In this section and the next, we make use of the positive definite matrix D^k, which we defined in (1.24) to be

$$D^k = (X^k)^{1/2}(S^k)^{-1/2}. \tag{6.24}$$

We also make repeated use of the matrix norm defined for a matrix $M \in \mathbb{R}^{p\times q}$ by

$$\|M\| \stackrel{\text{def}}{=} \max_{u\in\mathbb{R}^q:\|u\|=1} \|Mu\|,$$

a definition that holds for all three norms $\|\cdot\|_1$, $\|\cdot\|_2$, and $\|\cdot\|_\infty$. It is a simple consequence of this definition that

$$\|Mu\| \le \|M\|\|u\| \qquad \text{for all } u \in \mathbb{R}^q. \tag{6.25}$$

The next two lemmas give bounds on the scaled vectors $(D^k)^{-1}\Delta x^k$ and $D^k\Delta s^k$. As in the preceding section, the first result applies to any starting point (x^0,λ^0,s^0) for which $(x^0,s^0) > 0$, while the second result holds for the specific choice (6.12) and (6.13) and obtains a bound in which the dependence on n is explicit.

Lemma 6.5 *There is a positive constant C_2 such that*

$$\|(D^k)^{-1}\Delta x^k\| \le C_2\mu_k^{1/2}, \qquad \|D^k\Delta s^k\| \le C_2\mu_k^{1/2} \tag{6.26}$$

for all $k \ge 0$.

Proof. Let us define

$$(\bar{x},\bar{\lambda},\bar{s}) = (\Delta x,\Delta\lambda,\Delta s) + \nu_k(x^0,\lambda^0,s^0) - \nu_k(x^*,\lambda^*,s^*). \tag{6.27}$$

It is easy to verify that $(\bar{x},\bar{\lambda},\bar{s})$ satisfies (6.15), so from (6.16) we have

$$\left[\Delta x + \nu_k(x^0 - x^*)\right]^T \left[\Delta s + \nu_k(s^0 - s^*)\right] = 0. \tag{6.28}$$

From the last row in the system (6.4), we have

$$\begin{aligned}
& S(\Delta x + \nu_k(x^0 - x^*)) + X(\Delta s + \nu_k(s^0 - s^*)) \\
= \; & -XSe + \sigma\mu_k e + \nu_k S(x^0 - x^*) + \nu_k X(s^0 - s^*).
\end{aligned}$$

If we multiply this system by $(XS)^{-1/2}$ and note that $(XS)^{-1/2}S = D^{-1}$ and $(XS)^{-1/2}X = D$, we obtain

$$\begin{aligned}
& D^{-1}(\Delta x + \nu_k(x^0 - x^*)) + D(\Delta s + \nu_k(s^0 - s^*)) \qquad (6.29) \\
= \; & -(XS)^{-1/2}(XSe - \sigma\mu_k e) + \nu_k D^{-1}(x^0 - x^*) + \nu_k D(s^0 - s^*).
\end{aligned}$$

Now, because of (6.28), we have

$$\begin{aligned}
& \|D^{-1}(\Delta x + \nu_k(x^0 - x^*)) + D(\Delta s + \nu_k(s^0 - s^*))\|^2 \qquad (6.30) \\
= \; & \|D^{-1}(\Delta x + \nu_k(x^0 - x^*))\|^2 + \|D(\Delta s + \nu_k(s^0 - s^*))\|^2.
\end{aligned}$$

Taking squared norms of both sides in (6.29) and using (6.25) and (6.30), we obtain

$$\begin{aligned}
& \|D^{-1}(\Delta x + \nu_k(x^0 - x^*))\|^2 + \|D(\Delta s + \nu_k(s^0 - s^*))\|^2 \\
\leq \; & \left\{\|(XS)^{-1/2}\|\|XSe - \sigma\mu_k e\| + \nu_k\|D^{-1}(x^0 - x^*)\| + \nu_k\|D(s^0 - s^*)\|\right\}^2.
\end{aligned}$$

Let us now isolate the first term on the left-hand side and write

$$\begin{aligned}
& \|D^{-1}(\Delta x + \nu_k(x^0 - x^*))\| \\
\leq \; & \|(XS)^{-1/2}\|\|XSe - \sigma\mu_k e\| + \nu_k\|D^{-1}(x^0 - x^*)\| + \nu_k\|D(s^0 - s^*)\|.
\end{aligned}$$

A simple application of the triangle inequality and addition of an extra term $\nu_k\|D(s^0 - s^*)\|$ to the right-hand side give

$$\begin{aligned}
& \|D^{-1}\Delta x\| \qquad (6.31) \\
\leq \; & \|D^{-1}(\Delta x + \nu_k(x^0 - x^*))\| + \nu_k\|D^{-1}(x^0 - x^*)\| \\
\leq \; & \|(XS)^{-1/2}\|\|XSe - \sigma\mu_k e\| + 2\nu_k\|D^{-1}(x^0 - x^*)\| + 2\nu_k\|D(s^0 - s^*)\|.
\end{aligned}$$

We complete the proof by showing that each term on the right-hand side of (6.31) is $O(\mu_k^{1/2})$ in magnitude. For the first term, we have

$$\begin{aligned}
& \|XSe - \sigma\mu_k e\|^2 = \|XSe\|^2 - 2\sigma\mu_k x^T s + \sigma^2\mu_k^2 n \\
\leq \; & \|XSe\|_1^2 - 2\sigma n\mu_k^2 + \sigma^2 n\mu_k^2 \leq (n\mu_k)^2 - 2\sigma n\mu_k^2 + \sigma^2 n\mu_k^2 \leq n^2\mu_k^2,
\end{aligned}$$

while

$$\|(XS)^{-1/2}\| = \max_{i=1,\ldots,n} \frac{1}{(x_i s_i)^{1/2}} \leq \frac{1}{\gamma^{1/2}\mu_k^{1/2}}. \tag{6.32}$$

Hence, for the first term in (6.31), we obtain

$$\|(XS)^{-1/2}\|\|XSe - \sigma\mu_k e\| \leq \frac{n}{\gamma^{1/2}}\mu_k^{1/2}. \tag{6.33}$$

For the final two terms in (6.31), we have

$$\begin{aligned} &\nu_k\|D^{-1}(x^0 - x^*)\| + \nu_k\|D(s^0 - s^*)\| \\ \leq\ &\nu_k \left(\|D^{-1}\| + \|D\|\right) \max(\|x^0 - x^*\|, \|s^0 - s^*\|). \end{aligned} \tag{6.34}$$

For the matrix norm $\|D^{-1}\|$, we have

$$\|D^{-1}\| = \max_{i=1,\ldots,n} \|D_{ii}^{-1}\| = \|D^{-1}e\|_\infty = \|(XS)^{-1/2}Se\|_\infty \leq \|(XS)^{-1/2}\|\|s\|_1,$$

and similarly

$$\|D\| \leq \|(XS)^{-1/2}\|\|x\|_1.$$

From (6.34), (6.32), and Lemma 6.3, we have

$$\begin{aligned} &\nu_k\|D^{-1}(x^0 - x^*)\| + \nu_k\|D(s^0 - s^*)\| \\ \leq\ &\nu_k\|(x, s)\|_1\|(XS)^{-1/2}\| \max(\|x^0 - x^*\|, \|s^0 - s^*\|) \\ \leq\ &\frac{C_1}{\gamma^{1/2}}\mu_k^{1/2} \max(\|x^0 - x^*\|, \|s^0 - s^*\|). \end{aligned}$$

Finally, we obtain the bound on $\|D^{-1}\Delta x\|$ in (6.26) by combining the inequalities (6.31), (6.33), and (6.34) and defining C_2 as

$$C_2 = 2\frac{C_1}{\gamma^{1/2}} \max(\|x^0 - x^*\|, \|s^0 - s^*\|) + \frac{n}{\gamma^{1/2}}.$$

The other vector $D\Delta s$ also satisfies (6.31), so its bound is the same as the bound for $D^{-1}\Delta x$, and our proof is complete. □

A minor variation on Lemma 6.5 gives the bound for the starting point $(x^0, \lambda^0, s^0) = (\zeta e, 0, \zeta e)$.

Lemma 6.6 *Suppose that the starting point is chosen as in* (6.12) *and* (6.13). *Then there is a constant* ω *independent of* n *such that*

$$\|(D^k)^{-1}\Delta x^k\| \leq \omega n \mu_k^{1/2}, \qquad \|D^k \Delta s^k\| \leq \omega n \mu_k^{1/2}. \tag{6.35}$$

Proof. We proceed as in the proof of Lemma 6.5 to obtain (6.31) and (6.33). In place of (6.34), however, we obtain a more specialized bound. Because of the choice of ζ, we have

$$0 \le x^0 - x^* \le \zeta e, \qquad 0 \le s^0 - s^* \le \zeta e,$$

and therefore

$$\begin{aligned} \nu_k \|D^{-1}(x^0 - x^*)\| + \nu_k \|D(s^0 - s^*)\| &\le \nu_k \zeta \left[\|D^{-1}e\| + \|De\| \right] \\ &= \nu_k \zeta \left[\|(XS)^{-1/2}s\| + \|(XS)^{-1/2}x\| \right] \\ &\le \nu_k \zeta \|(XS)^{-1/2}\| \|(x,s)\|_1. \end{aligned}$$

Hence, from (6.32) and (6.23), we have

$$\nu_k \|D^{-1}(x^0 - x^*)\| + \nu_k \|D(s^0 - s^*)\| \le \frac{1}{\gamma^{1/2}\mu_k^{1/2}} 4\beta n \mu_k = \frac{4\beta}{\gamma^{1/2}} n \mu_k^{1/2}. \tag{6.36}$$

The result is obtained from (6.31), (6.33), and (6.36) by setting

$$\omega = \frac{9\beta}{\gamma^{1/2}}. \qquad \Box$$

Technical Results III: A Uniform Lower Bound on α_k

Lemmas 6.3–6.6 set the scene for the following crucial result: There is a value $\bar{\alpha} > 0$ such that the conditions (6.5) and (6.6) hold for all $\alpha \in [0, \bar{\alpha}]$ at all iterates k. As we will see, this property guarantees that Algorithm IPF can make significant progress at every step.

Lemma 6.7 *There is a value $\bar{\alpha} \in (0,1)$ such that the following three conditions are satisfied for all $\alpha \in [0, \bar{\alpha}]$ and all $k \ge 0$:*

$$\begin{aligned} (x^k + \alpha \Delta x^k)^T (s^k + \alpha \Delta s^k) &\ge (1-\alpha)(x^k)^T s^k, & \text{(6.37a)} \\ (x_i^k + \alpha \Delta x_i^k)(s_i^k + \alpha \Delta s_i^k) &\ge (\gamma/n)(x^k + \alpha \Delta x^k)^T (s^k + \alpha \Delta s^k), & \text{(6.37b)} \\ (x^k + \alpha \Delta x^k)^T (s^k + \alpha \Delta s^k) &\le (1 - .01\alpha)(x^k)^T s^k. & \text{(6.37c)} \end{aligned}$$

Therefore, the conditions (6.5) *and* (6.6) *are satisfied for all $\alpha \in [0, \bar{\alpha}]$ and all $k \ge 0$.*

If the conditions of Lemma 6.6 *are satisfied, then*

$$\bar{\alpha} \ge \bar{\delta}/n^2 \tag{6.38}$$

for some positive scalar $\bar{\delta}$ independent of n.

Proof. Note first that, because of (6.26), we have

$$(\Delta x)^T \Delta s = (D^{-1}\Delta x)^T(D\Delta s) \leq \|D^{-1}\Delta x\|\|D\Delta s\| \leq C_2^2\mu_k. \qquad (6.39)$$

Similarly, we have

$$|\Delta x_i \Delta s_i| = |D_{ii}^{-1}\Delta x_i||D_{ii}\Delta s_i| \leq \|D^{-1}\Delta x\|\|D\Delta s\| \leq C_2^2\mu_k. \qquad (6.40)$$

From the last row of the system (6.4), we deduce two more useful equalities. By summing over all n components of this equation, we obtain

$$s^T\Delta x + x^T\Delta s = e^T(S\Delta x + X\Delta s) = e^T(-XSe + \sigma_k\mu e) = (\sigma_k - 1)x^T s, \quad (6.41)$$

whereas by taking a single component, we obtain

$$s_i\Delta x_i + x_i\Delta s_i = -x_i s_i + \sigma_k\mu. \qquad (6.42)$$

For each of the three inequalities in (6.37) in turn, we derive conditions on $\bar{\alpha}$ under which these bounds are guaranteed to hold. For (6.37a), we have

$$\begin{aligned} &(x + \alpha\Delta x)^T(s + \alpha\Delta s) \\ &= x^T s + \alpha(\sigma_k - 1)x^T s + \alpha^2 \Delta x^T \Delta s \\ &\geq (1 - \alpha)x^T s + \alpha\sigma_k x^T s - \alpha^2 C_2^2 \mu_k \\ &\geq (1 - \alpha)x^T s + \left(\alpha\sigma_{\min} - \alpha^2 C_2^2/n\right) x^T s. \end{aligned} \qquad (6.43)$$

Hence, (6.37a) holds if the last term in (6.43) is nonnegative, which is true if α satisfies

$$\alpha \leq \frac{n\sigma_{\min}}{C_2^2}. \qquad (6.44)$$

Note that (6.37a) implies that the first condition in (6.2) is satisfied, since for

$$r_b^k(\alpha) = b - Ax^k(\alpha), \qquad r_c^k(\alpha) = A^T\lambda^k(\alpha) + s^k(\alpha) - c,$$

we have

$$\frac{\|(r_b^k(\alpha), r_c^k(\alpha))\|}{\mu_k(\alpha)} \leq \frac{(1-\alpha)\|(r_b^k, r_c^k)\|}{\mu_k(\alpha)} \leq \frac{\|(r_b^k, r_c^k)\|}{\mu_k} \leq \beta\frac{\|(r_b^0, r_c^0)\|}{\mu_0}.$$

For (6.37b), we use (6.40) and (6.42) and the fact that $x_i s_i \geq \gamma\mu_k$ to write

$$\begin{aligned} (x_i + \alpha\Delta x_i)(s_i + \alpha\Delta s_i) &\geq x_i s_i(1 - \alpha) + \alpha\sigma_k\mu_k - \alpha^2 C_2^2\mu_k \\ &\geq \gamma(1 - \alpha)\mu_k + \alpha\sigma_k\mu_k - \alpha^2 C_2^2\mu_k. \end{aligned} \qquad (6.45)$$

On the other hand, we can show as in (6.43) that

$$\frac{1}{n}(x+\alpha\Delta x)^T(s+\alpha\Delta s) \le (1-\alpha)\mu_k + \alpha\sigma_k\mu_k + \alpha^2 C_2^2\mu_k/n. \tag{6.46}$$

Hence, by taking the differences of the two sides in (6.37b) and using (6.45) and (6.46), we obtain

$$\begin{aligned}
&(x_i+\alpha\Delta x_i)(s_i+\alpha\Delta s_i) - (\gamma/n)(x+\alpha\Delta x)^T(s+\alpha\Delta s) \\
&\quad \ge \alpha\sigma_k(1-\gamma)\mu_k - (1+\gamma/n)\alpha^2 C_2^2\mu_k \ge \alpha\sigma_{\min}(1-\gamma)\mu_k - 2\alpha^2 C_2^2\mu_k.
\end{aligned}$$

The bound (6.37b) holds if the final expression is nonnegative, which is true if α satisfies

$$\alpha \le \frac{\sigma_{\min}(1-\gamma)}{2C_2^2}. \tag{6.47}$$

Finally, we take the differences of the two sides in (6.37c) and use (6.46) and $\sigma_k \le \sigma_{\max} \le 0.5$ to write

$$\begin{aligned}
&\frac{1}{n}(x+\alpha\Delta x)^T(s+\alpha\Delta s) - (1-0.01\alpha)\mu_k \\
&\quad \le (1-\alpha)\mu_k + \alpha\sigma_k\mu_k + \alpha^2 C_2^2\mu_k/n - (1-0.01\alpha)\mu_k \\
&\quad \le -0.99\alpha\mu_k + 0.5\alpha\mu_k + \alpha^2 C_2^2\mu_k/n \\
&\quad = -0.49\alpha\mu_k + \alpha^2 C_2^2\mu_k/n.
\end{aligned}$$

If α satisfies the bound

$$\alpha \le \frac{0.49n}{C_2^2}, \tag{6.48}$$

the last expression is nonpositive, and therefore (6.37c) holds.

Combining the bounds (6.44), (6.47), and (6.48), we conclude that the conditions (6.37) all hold if $\alpha \in [0, \bar{\alpha}]$, where

$$\bar{\alpha} = \min\left(\frac{n\sigma_{\min}}{C_2^2}, \frac{\sigma_{\min}(1-\gamma)}{C_2^2}, \frac{0.49n}{C_2^2}, 1\right).$$

The second part of the result—the bound (6.38) for the special starting point—is obtained from minor modifications to the analysis above. We leave this part of the proof to the exercises. □

Proofs of Theorems 6.1 and 6.2

The main results of the chapter follow easily from Lemma 6.7. First, we prove the result concerning global convergence from any starting point $(x^0, \lambda^0, s^0) \in \mathcal{N}_{-\infty}(\gamma, \beta)$.

Proof of Theorem 6.1. Q-linear convergence of $\{\mu_k\}$ follows directly from (6.10) and (6.38). For the residuals, we have from (6.2) that

$$\|(r_b^k, r_c^k)\| \leq \mu_k \beta \frac{\|(r_b^0, r_c^0)\|}{\mu_0}.$$

Therefore, the sequence of residuals norms is bounded above by another sequence that converges Q-linearly, so $\{\|(r_b^k, r_c^k)\|\}$ is R-linearly convergent. □

Finally, we prove the polynomial complexity result, which is a special case of Theorem 3.2 from Chapter 3.

Proof of Theorem 6.2. Note from Theorem 6.1 and (6.38) that

$$\mu_{k+1} \leq (1 - .01\bar{\alpha})\mu_k \leq (1 - .01\bar{\delta}/n^2)\mu_k, \qquad k = 0, 1, \ldots,$$

so that (3.10) holds with $\delta = .01\bar{\delta}$ and $\omega = 2$. The result now follows immediately from Theorem 3.2. □

Limit Points of the Iteration Sequence

In Chapter 5, we showed that the sequences of x^k and s^k components generated by path-following algorithms are bounded and that their limit points can be used to construct strictly complementary primal-dual solutions to the linear programming problem. For Algorithm IPF of this chapter, we show in the first result below, Theorem 6.8, that limit points correspond to strictly complementary primal-dual solutions. The second result, Theorem 6.9, concerns boundedness of the sequence $\{(x^k, s^k)\}$.

For the following results, we recall the partitioning $\mathcal{B} \cup \mathcal{N} = \{1, 2, \ldots, n\}$ from (2.11).

Theorem 6.8 *Let $\{(x^k, \lambda^k, s^k)\}$ be the sequence of iterates generated by Algorithm IPF. Then there is a constant C_3 such that for all k sufficiently large, we have*

$$0 < x_i^k \leq \mu_k / C_3 \;\; (i \in \mathcal{N}), \qquad 0 < s_i^k \leq \mu_k / C_3 \;\; (i \in \mathcal{B}), \tag{6.49}$$

$$s_i^k \geq C_3 \gamma \;\; (i \in \mathcal{N}), \qquad x_i^k \geq C_3 \gamma \;\; (i \in \mathcal{B}). \tag{6.50}$$

Hence, any limit point of the sequence $\{(x^k, s^k)\}$ can be used to construct a strictly complementary solution of (2.1), (2.2).

Proof. From Theorem 6.1 and (6.8), we have $\nu_k \downarrow 0$, so we can define an index $K_{0.5}$ such that $\nu_k \leq 0.5$ for all $k \geq K_{0.5}$. We now find a value of C_3 for which (6.49) and (6.50) hold for all $k \geq K_{0.5}$.

The expression (6.19) turns out to be useful once again. Let (x^*, λ^*, s^*) be any strictly complementary solution, and note that

$$(x^0)^T s + x^T s^0 > 0, \qquad (x^*)^T s^* = 0.$$

(As usual, we drop the index k from (x^k, λ^k, s^k).) We now rearrange (6.19) to obtain

$$\begin{aligned} x^T s^* + s^T x^* &\le \frac{\nu_k^2}{1-\nu_k} n\mu_0 + \frac{n\mu_k}{1-\nu_k} + \nu_k \left((x^0)^T s^* + (x^*)^T s^0 \right) \\ &\le 2\nu_k^2 n\mu_0 + 2n\mu_k + \nu_k \left((x^0)^T s^* + (x^*)^T s^0 \right), \end{aligned} \qquad (6.51)$$

where the second inequality follows from the fact that $1 - \nu_k \ge 0.5$ for $k \ge K_{0.5}$. Because of (6.9), we can bound ν_k by a multiple of μ_k, so from (6.51), there is a constant $\bar{C}_3 > 0$ such that

$$x^T s^* + s^T x^* \le \mu_k / \bar{C}_3.$$

As in Lemma 5.13, we obtain for $i \in \mathcal{N}$ with $s_i^* > 0$ that

$$x_i s_i^* \le \mu_k / \bar{C}_3 \;\Rightarrow\; x_i \le \frac{1}{\bar{C}_3 s_i^*} \mu_k.$$

A similar bound can be found for $s_i, i \in \mathcal{B}$. Therefore, the result (6.49) holds when we define

$$C_3 = \bar{C}_s \min \left(\min_{i \in \mathcal{N}} s_i^*, \min_{i \in \mathcal{B}} x_i^* \right). \qquad (6.52)$$

Since the iterates generated by Algorithm IPF satisfy $x_i s_i \ge \gamma\mu_k$, we have from (6.49) that

$$x_i \ge \frac{\gamma\mu_k}{s_i} \ge C_3\gamma \;\; (i \in \mathcal{B}), \qquad s_i \ge \frac{\gamma\mu_k}{x_i} \ge C_3\gamma \;\; (i \in \mathcal{N}),$$

proving (6.50).

To prove the final statement, let $\mathcal{K}$ be a subsequence such that

$$\lim_{k \in \mathcal{K}} (x^k, s^k) = (x^*, s^*).$$

By taking limits, we clearly have

$$Ax^* = b, \quad (x^*, s^*) \ge 0, \quad (x^*)^T s^* = 0.$$

Strict complementarity follows immediately from (6.50). For the remaining condition, we have

$$\lim_{k\in\mathcal{K}} r_c^k = \lim_{k\in\mathcal{K}} s^k - c + A^T\lambda^k = 0,$$

and therefore

$$\text{dist}(s^* - c, \text{Range}(A^T)) = 0.$$

Since Range(A^T) is a closed subspace, it follows that $s^* - c \in \text{Range}(A^T)$ and therefore there exists a vector λ^* such that $s^* - c + A^T\lambda^* = 0$. Therefore, (x^*, λ^*, s^*) is a strictly complementary solution. □

The definition (6.52) of C_3 differs from the definition (5.32) for the feasible case in the presence of the term $\bar{C}_3$. In the feasible case we have simply $\bar{C}_3 = n$, while in the infeasible case this quantity depends on (x^0, λ^0, s^0) and (x^*, λ^*, s^*).

To show that the sequence $\{(x^k, \lambda^k, s^k)\}$ is bounded, we need the additional assumption that the set of strictly feasible points $\mathcal{F}^o$ is nonempty. Algorithm IPF is still defined and still satisfies Theorems 6.1 and 6.2 if this assumption does not hold, while the algorithms of Chapter 5 are not defined when $\mathcal{F}^o = \emptyset$.

Theorem 6.9 *Let $\{(x^k, \lambda^k, s^k)\}$ be the sequence of iterates generated by Algorithm IPF, and suppose that $\mathcal{F}^o \neq \emptyset$. Then the partial sequence $\{(x^k, s^k)\}$ is bounded and therefore has at least one limit point.*

Proof. Since $\mathcal{F}^o \neq \emptyset$, we can choose a point $(\bar{x}, \bar{\lambda}, \bar{s})$ such that

$$A\bar{x} = b, \qquad A^T\bar{\lambda} + \bar{s} = c, \qquad (\bar{x}, \bar{s}) > 0.$$

Combining this definition with (6.1), we obtain

$$\begin{aligned}
(x^k - \bar{x})^T(\bar{s} - s^k) &= (x^k - \bar{x})^T(A^T\lambda^k - r_c^k - A^T\bar{\lambda}) \\
&= \left[A(x^k - \bar{x})\right]^T(\lambda^k - \bar{\lambda}) - (r_c^k)^T(x^k - \bar{x}) \\
&= (r_b^k)^T(\lambda^k - \bar{\lambda}) - (r_c^k)^T(x^k - \bar{x}). \qquad (6.53)
\end{aligned}$$

Now, since $r_b^k \in \text{Range}(A)$, we have by Hoffman's lemma (Lemma A.3 in Appendix A) that there is a constant $\hat{C}$ independent of k such that

$$r_b^k = Ax_b^k \qquad \text{for some } x_b^k \text{ with } \|x_b^k\| \le \hat{C}\|r_b^k\|.$$

Hence, the first term in (6.53) can be bounded by

$$\begin{aligned}
(r_b^k)^T(\bar{\lambda} - \lambda^k) &= (x_b^k)^T A^T(\bar{\lambda} - \lambda^k) \\
&= (x_b^k)^T(-\bar{s} + s^k - r_c^k) \le \hat{C}\|r_b^k\|(\|\bar{s}\| + \|s^k\| + \|r_c^k\|).
\end{aligned}$$

Using this bound together with (6.8), we find that

$$\begin{aligned}
&(r_b^k)^T(\lambda^k - \bar{\lambda}) - (r_c^k)^T(x^k - \bar{x}) \\
&\leq \hat{C}\nu_k\|r_b^0\|(\|s^k\| + \|\bar{s}\| + \|r_c^0\|) + \nu_k\|r_c^0\|(\|x^k\| + \|\bar{x}\|) \\
&\leq C_4\nu_k(\|(x^k, s^k)\| + 1)
\end{aligned}$$

for some appropriately defined constant C_4. Rearranging (6.53), we obtain

$$\bar{x}^T s^k + \bar{s}^T x^k \leq \bar{x}^T \bar{s} + n\mu_k + C_4\nu_k \left(\|(x^k, s^k)\| + 1\right). \tag{6.54}$$

By Lemma 6.3, the product $\nu_k\|(x^k, s^k)\|$ is $O(\mu_k)$, while ν_k itself is also $O(\mu_k)$ because of (6.2). Hence, the right-hand side of (6.54) is bounded above by $\bar{x}^T\bar{s} + C_5\mu_k$ for some constant C_5. Since $(\bar{x}, \bar{s}) > 0$, it follows from (6.54) that all components of (x^k, s^k) are bounded above by

$$\frac{1}{\min_{i=1,2,\ldots,n} \min(\bar{x}_i,\, \bar{s}_i)} (\bar{x}^T\bar{s} + C_5\mu_0). \quad \square$$

Exercises

1. Consider the following linear program in nonstandard form:

$$\min_x c^T x \qquad \text{subject to } Ax \leq b,\ Cx = d.$$

   Transforming the constraints and variables via the techniques described at the start of Chapter 2, reformulate this problem in (i) standard form (2.1) and (ii) dual form (2.2). Show that the second formulation has no strictly feasible points.

2. Algorithm IPF requires us to enforce a sufficient decrease criterion (6.6) on the step length α. Why is this criterion not required by Algorithm LPF?

3. In Algorithm IPF, we choose α_k to be the *largest* value of α for which the conditions (6.5) and (6.6) are satisfied. The same convergence properties hold, however, if we use the following more flexible choice of α_k:

   > α_k satisfies the conditions (6.5) and (6.6), and we have $\alpha_k \geq \bar{\alpha}_k/10$, where $\bar{\alpha}_k$ is the largest value of α for which these two conditions hold.

Outline the changes that are needed in the proofs of Theorems 6.1 and 6.2 to accommodate this choice of α_k.

4. Prove the second part of Lemma 6.7, that is, the bound (6.38) for the special starting point (6.12) and (6.13).

Chapter 7

Superlinear Convergence and Finite Termination

In all the algorithms we have discussed so far, the duality measures μ_k approach zero as $k \to \infty$ (provided that the solution set Ω is nonempty), and all limit points of the sequence $\{(x^k, s^k)\}$ correspond to primal-dual solutions. During the final stages of these algorithms, as the iterates approach Ω, the danger of Newton-like steps straying outside the feasible region diminishes. Hence, we can relax some of the restrictions on the search directions and take a step that is closer to a pure Newton step. Superlinearly convergent algorithms can be designed in this way. Alternatively, rather than continuing to iterate, we can jump directly to a solution from a sufficiently advanced point (x^k, λ^k, s^k), a process known as *finite termination*. Superlinear convergence and finite termination are closely related strategies. In this chapter, we describe them both and examine their relationship to each other.

Superlinear convergence analysis revolves around the relationship that all primal-dual algorithms have with Newton's method for nonlinear equations, but the results have a slightly different character from the standard Kantorovich analysis of Newton-like methods [56]. Kantorovich essentially shows that if z^* is a solution of $F(z) = 0$ at which F is smooth and the Jacobian $J(z^*)$ is nonsingular, the Newton iterates z^k converge quadratically to z^*, provided that the starting point is close enough to this solution. Our function F defined by (1.4) is only mildly nonlinear, and some variables are constrained to be nonnegative. These features allow us to prove surprisingly strong results. Superlinear convergence occurs for all feasible linear programming problems, even when the limiting Jacobian is singular and the solution is nonunique.

Although the steps taken by a superlinearly convergent algorithm become more and more similar to pure Newton steps, we cannot simply start to take pure Newton steps at some point in the algorithm. The safeguards of Chapters 4, 5, and 6 must be maintained, albeit in relaxed form, to ensure global convergence.

The main technical result of this chapter is that the Δx and Δs components of the affine-scaling step—the pure Newton step calculated by setting $\sigma = 0$ in (1.12) and (1.20)—have size $O(\mu)$. Superlinearity of Algorithm PC from Chapter 5 follows as an immediate consequence of this result. We show, too, that the long-step path-following method, Algorithm LPF, can be accelerated so that its sequence of duality measures $\{\mu_k\}$ converges almost quadratically to zero. The advantage of the latter approach over Algorithm PC is that it requires just one factorization of the matrix in (1.12), (1.20) at each step, instead of two.

After discussing superlinear convergence, we turn to finite termination. We describe a procedure due to Ye [155]—Procedure FT—that converts any sufficiently advanced iterate (x^k, λ^k, s^k) of a primal-dual algorithm to an exact solution $(x^*, \lambda^*, s^*) \in \Omega$.

Our analysis in this chapter makes extensive use of the partition

$$\mathcal{B} \cup \mathcal{N} = \{1, 2, \ldots, n\}$$

that was defined in (2.11). The data-dependent quantity $\epsilon(A, b, c)$, defined in (3.5) and used repeatedly in Chapters 3, 5, and 6, is also central to our analysis. We see in this chapter that superlinear convergence and finite termination become effective only when μ_k falls below a threshold that depends on $\epsilon(A, b, c)$.

Affine-Scaling Steps

We examine the asymptotic behavior of primal-dual affine-scaling steps, which are calculated by setting $\sigma = 0$ in (1.12) and (1.20). We have seen already that search directions of this type are used in Algorithm PC of Chapter 5, the predictor-corrector method. Other path-following methods, including Algorithm LPF of Chapter 5 and Algorithm IPF of Chapter 6, can also be modified to use affine-scaling search directions at later iterations. Used judiciously, these steps can yield superlinear convergence, instead of the linear convergence rates that were proven in the earlier chapters.

For $(x, \lambda, s) \in \mathcal{F}^o$, the affine-scaling direction $(\Delta x, \Delta\lambda, \Delta s)$ satisfies

$$\begin{bmatrix} 0 & A^T & I \\ A & 0 & 0 \\ S & 0 & X \end{bmatrix} \begin{bmatrix} \Delta x \\ \Delta\lambda \\ \Delta s \end{bmatrix} = \begin{bmatrix} 0 \\ 0 \\ -XSe \end{bmatrix}. \tag{7.1}$$

The generic affine-scaling step equation for infeasible-interior-point algorithms is

$$\begin{bmatrix} 0 & A^T & I \\ A & 0 & 0 \\ S & 0 & X \end{bmatrix} \begin{bmatrix} \Delta x \\ \Delta\lambda \\ \Delta s \end{bmatrix} = \begin{bmatrix} -r_c \\ -r_b \\ -XSe \end{bmatrix}, \tag{7.2}$$

where the residuals r_b and r_c are defined as usual by

$$r_b = Ax - b, \qquad r_c = A^T\lambda + s - c.$$

In the next section, we show that the Δx and Δs components of affine-scaling steps are $O(\mu)$ in size, under the assumption that (x, λ, s) lies in a certain neighborhood of the central path $\mathcal{C}$. When these estimates hold, the linear equations (7.1) and (7.2) are close approximations to the nonlinear equations (1.4a), (1.4b), and (1.4c), which, along with the positivity conditions (1.4d), define the solution set Ω. We can thus take a step of length very close to 1 along the direction $(\Delta x, \Delta\lambda, \Delta s)$, producing a new iterate (x^+, λ^+, s^+) that is much closer to the solution set than the current point (x, λ, s).

In the Kantorovich theory for Newton's method [56], it is easy to show that the step has similar magnitude to the current function vector F. The Jacobian (the coefficient matrix in the step equations) approaches a nonsingular matrix, so the Newton step satisfies

$$-J(z)^{-1}F(z) \approx -J(z^*)^{-1}F(z) = O(\|F(z)\|).$$

In our case, the coefficient matrix in (7.1) and (7.2) approaches a nonsingular limit only in the nondegenerate case, in which Ω contains a single point. To obtain the estimate of $(\Delta x, \Delta\lambda, \Delta s)$ for the general case in which $\mathcal{B}$ and $\mathcal{N}$ contain any number of elements between 0 and n, some highly technical analysis is needed. Some of the necessary groundwork has already been laid in Chapter 6.

An Estimate of $(\Delta x, \Delta s)$: The Feasible Case

It is relatively easy to find bounds on half the required components, namely, $\Delta x_{\mathcal{N}}$ and $\Delta s_{\mathcal{B}}$. We first obtain this bound for the feasible case

(7.1) under the assumption that the current point (x, λ, s) belongs to the neighborhood $\mathcal{N}_{-\infty}(\gamma)$.

Lemma 7.1 *Suppose that $(x, \lambda, s) \in \mathcal{N}_{-\infty}(\gamma)$ and that the affine-scaling direction $(\Delta x, \Delta\lambda, \Delta s)$ is calculated from (7.1). Then there is a constant C_4 such that*

$$\|\Delta x_{\mathcal{N}}\| \le C_4\mu, \qquad \|\Delta s_{\mathcal{B}}\| \le C_4\mu.$$

Proof. Since the conditions of Lemma 5.13 are satisfied, we know that there is a positive constant C_3 such that

$$0 < x_i \le \mu/C_3 \;\; (i \in \mathcal{N}), \qquad 0 < s_i \le \mu/C_3 \;\; (i \in \mathcal{B}). \tag{7.3}$$

In fact, by comparing the definition (5.32) of C_3 with our definition (3.5) of $\epsilon(A, b, c)$, we see immediately that

$$C_3 = \frac{\epsilon(A, b, c)}{n}. \tag{7.4}$$

Let us recall the definition of the positive diagonal matrix $D = X^{1/2}S^{-1/2}$ from (1.24). Multiplying the last block row in (7.1) by the diagonal matrix $(XS)^{-1/2}$, we obtain

$$D^{-1}\Delta x + D\Delta s = -(XS)^{1/2}e.$$

Taking inner products of both sides of this expression, we obtain

$$\|D^{-1}\Delta x\|^2 + 2\Delta x^T\Delta s + \|D\Delta s\|^2 = \|(XS)^{1/2}e\|^2 = x^Ts = n\mu.$$

Because $\Delta x^T\Delta s = 0$ (Lemma 5.1) and the other two terms on the left-hand side of this expression are positive, we can write

$$\|D^{-1}\Delta x\|^2 \le n\mu, \qquad \|D\Delta s\|^2 \le n\mu. \tag{7.5}$$

Noting that the ith diagonal element of D is $\sqrt{x_i/s_i}$, we have for any i that

$$\frac{s_i}{x_i}(\Delta x_i)^2 \le \|D^{-1}\Delta x\|^2 \le n\mu.$$

Hence, choosing $i \in \mathcal{N}$ and using $x_is_i \ge \gamma\mu$ and (7.3), we find that

$$(\Delta x_i)^2 \le \frac{n\mu x_i}{s_i} = \frac{n\mu x_i^2}{x_is_i} \le \frac{n\mu\mu^2}{\gamma\mu C_3^2} = \frac{n}{\gamma C_3^2}\mu^2.$$

Therefore, we have

$$\|\Delta x_{\mathcal{N}}\| \leq \sqrt{n} \max_{i \in \mathcal{N}} |\Delta x_i| \leq \frac{n}{\gamma^{1/2} C_3} \mu,$$

and the bound on $\|\Delta x_{\mathcal{N}}\|$ follows if we define

$$C_4 = \frac{n}{\gamma^{1/2} C_3}. \tag{7.6}$$

The argument for $\|\Delta s_{\mathcal{B}}\|$ is identical. □

To find bounds on the remaining components $\Delta x_{\mathcal{B}}$ and $\Delta s_{\mathcal{N}}$, we formulate two weighted least-squares problems for which these vectors are the solutions. The following technical result is then used to show that both vectors are $O(\mu)$. Its proof can be found in Appendix A.

Lemma 7.2 *Let the matrix $H \in \mathbb{R}^{p \times q}$ be given. Then there exists a nonnegative constant $\hat{C}$ depending only on H with the following property: For any vector $h \in$ Range(H) and any nonsingular diagonal matrix Σ, the unique solution $\bar{w}$ of the problem*

$$\min_w \tfrac{1}{2} \|\Sigma w\|^2 \quad \textit{subject to } Hw = h \tag{7.7}$$

satisfies

$$\|\bar{w}\| \leq \hat{C} \|h\|.$$

The next lemma is due essentially to Adler and Monteiro [1]. For notation, we use $A_{\mathcal{B}}$ and $A_{\mathcal{N}}$ to denote a column partition of the matrix A according to the sets $\mathcal{B}$ and $\mathcal{N}$, and $D_{\mathcal{B}}$ and $D_{\mathcal{N}}$ to denote diagonal submatrices of D obtained from the same partition.

Lemma 7.3 *Suppose that the assumptions of Lemma* 7.1 *hold. Then*

(i) *$u = \Delta x_{\mathcal{B}}$ is the unique solution of the following convex quadratic programming problem:*

$$\min_u \tfrac{1}{2} \|D_{\mathcal{B}}^{-1} u\|^2 \quad \textit{subject to} \quad A_{\mathcal{B}} u = -A_{\mathcal{N}} \Delta x_{\mathcal{N}}. \tag{7.8}$$

(ii) *$(v, \pi) = (\Delta s_{\mathcal{N}}, \Delta \lambda)$ is a solution of the following convex quadratic programming problem:*

$$\min_{(v,\pi)} \tfrac{1}{2} \|D_{\mathcal{N}} v\|^2 \quad \textit{subject to} \quad \begin{array}{l} A_{\mathcal{B}}^T \pi = -\Delta s_{\mathcal{B}}, \\ A_{\mathcal{N}}^T \pi + v = 0. \end{array} \tag{7.9}$$

The $\Delta s_{\mathcal{N}}$ component is unique.

Proof. Our result about sufficiency of the KKT conditions, Theorem A.2(i), says that $u = \Delta x_{\mathcal{B}}$ solves (7.8) if there exists a vector $\hat{\pi}$ such that

$$\begin{aligned} D_{\mathcal{B}}^{-2}\Delta x_{\mathcal{B}} - A_{\mathcal{B}}^T\hat{\pi} &= 0, & \text{(7.10a)} \\ A_{\mathcal{B}}\Delta x_{\mathcal{B}} &= -A_{\mathcal{N}}\Delta x_{\mathcal{N}}. & \text{(7.10b)} \end{aligned}$$

The second equation, (7.10b), is satisfied because of (7.1). Combining the first and third block rows in (7.1) and examining the $\mathcal{B}$ indices only, we have

$$-D_{\mathcal{B}}^{-2}\Delta x_{\mathcal{B}} + A_{\mathcal{B}}^T\Delta\lambda = s_{\mathcal{B}}. \tag{7.11}$$

Since (x, λ, s) is feasible, we have $s_{\mathcal{B}} = -A_{\mathcal{B}}^T\lambda + c_{\mathcal{B}}$, whereas for any solution (x^*, λ^*, s^*), we have $c_{\mathcal{B}} = A_{\mathcal{B}}^T\lambda^*$. Hence from (7.11), we have

$$-D_{\mathcal{B}}^{-2}\Delta x_{\mathcal{B}} + A_{\mathcal{B}}^T\Delta\lambda = -A_{\mathcal{B}}^T\lambda + c_{\mathcal{B}} = -A_{\mathcal{B}}^T(\lambda - \lambda^*).$$

Hence, (7.10a) holds when we set

$$\hat{\pi} = \lambda + \Delta\lambda - \lambda^*.$$

Uniqueness follows directly from Theorem A.2(ii).

For (ii), we appeal again to Theorem A.2(i), which tells us that $(v, \pi) = (\Delta s_{\mathcal{N}}, \Delta\lambda)$ solves (7.9) if there are vectors $\hat{u}_{\mathcal{B}}$ and $\hat{u}_{\mathcal{N}}$ such that

$$\begin{aligned} D_{\mathcal{N}}^2\Delta s_{\mathcal{N}} - \hat{u}_{\mathcal{N}} &= 0, & \text{(7.12a)} \\ -A_{\mathcal{B}}\hat{u}_{\mathcal{B}} - A_{\mathcal{N}}\hat{u}_{\mathcal{N}} &= 0, & \text{(7.12b)} \\ A_{\mathcal{B}}^T\Delta\lambda &= -\Delta s_{\mathcal{B}}, & \text{(7.12c)} \\ A_{\mathcal{N}}^T\Delta\lambda + \Delta s_{\mathcal{N}} &= 0. & \text{(7.12d)} \end{aligned}$$

The last two equations follow immediately from the first block row of (7.1). By manipulating (7.1), it is also easy to show that the following choices for $\hat{u}_{\mathcal{N}}$ and $\hat{u}_{\mathcal{B}}$ satisfy (7.12a) and (7.12b):

$$\hat{u}_{\mathcal{N}} = x_{\mathcal{N}} + \Delta x_{\mathcal{N}}, \qquad \hat{u}_{\mathcal{B}} = x_{\mathcal{B}} + \Delta x_{\mathcal{B}} - x_{\mathcal{B}}^*, \tag{7.13}$$

where x^* is any primal solution.

We prove uniqueness of the $\Delta s_{\mathcal{N}}$ component only for the case in which A has full row rank, and leave the general case as an exercise. If we multiply the first constraint in (7.9) by $A_{\mathcal{B}}$ and the second by $A_{\mathcal{N}}$ and add, we obtain

$$(A_{\mathcal{B}}A_{\mathcal{B}}^T + A_{\mathcal{N}}A_{\mathcal{N}}^T)\pi + A_{\mathcal{N}}v = AA^T\pi + A_{\mathcal{N}}v = -A_{\mathcal{B}}\Delta s_{\mathcal{B}}.$$

Since A has full row rank, we can write

$$\pi = -(AA^T)^{-1}(A_{\mathcal{N}}v + A_{\mathcal{B}}\Delta s_{\mathcal{B}}), \tag{7.14}$$

so (7.9) can be reformulated as

$$\begin{aligned} \min_v \tfrac{1}{2}\|D_{\mathcal{N}}v\|^2 \text{ subject to} \\ (I - A_{\mathcal{N}}^T(A^TA)^{-1}A_{\mathcal{N}})v = A_{\mathcal{N}}^T(A^TA)^{-1}A_{\mathcal{B}}\Delta s_{\mathcal{B}}. \end{aligned} \tag{7.15}$$

Since this problem has a strictly convex objective, uniqueness of its solution $v = \Delta s_{\mathcal{N}}$ follows from Theorem A.2(ii). We recover $\pi = \Delta\lambda$ by substituting in (7.14). □

Theorem 7.4 *Given* $(x, \lambda, s) \in \mathcal{N}_{-\infty}(\gamma)$ *and* $(\Delta x, \Delta\lambda, \Delta s)$ *computed from* (7.1), *there is a constant* C_5 *such that*

$$\|(\Delta x, \Delta s)\| \le C_5\mu.$$

Proof. We proved in Lemma 7.1 that $\|\Delta x_{\mathcal{N}}\|$ and $\|\Delta s_{\mathcal{B}}\|$ are bounded by $C_4\mu$, so it remains to show that $\|\Delta x_{\mathcal{B}}\|$ and $\|\Delta s_{\mathcal{N}}\|$ are also $O(\mu)$. Since the diagonals of $D_{\mathcal{B}}$ are strictly positive, we can apply Lemma 7.2 directly to the convex quadratic program (7.8). Hence, there is a constant $\hat{C}_1$ depending solely on $A_{\mathcal{B}}$ such that

$$\|\Delta x_{\mathcal{B}}\| \le \hat{C}_1\|A_{\mathcal{N}}\Delta x_{\mathcal{N}}\| \le \hat{C}_1\|A_{\mathcal{N}}\|C_4\mu,$$

where the last inequality follows from Lemma 7.1. Similarly, from (7.15), there is a constant $\hat{C}_2$ depending solely on A such that

$$\|\Delta s_{\mathcal{N}}\| \le \hat{C}_2\|A_{\mathcal{N}}^T(A^TA)^{-1}A_{\mathcal{B}}\|\|\Delta s_{\mathcal{B}}\| \le \hat{C}_2\|A_{\mathcal{N}}^T(A^TA)^{-1}A_{\mathcal{B}}\|C_4\mu.$$

The result follows when we define C_5 in an obvious way. □

An Estimate of $(\Delta x, \Delta s)$: The Infeasible Case

A result similar to Theorem 7.4 holds for the steps of an infeasible-interior-point algorithm. The result applies specifically to the algorithm of Chapter 6, but it is also useful in other contexts, so we state it in general terms here.

Theorem 7.5 *Suppose that the iteration sequence* (x^k, λ^k, s^k) *is generated by an infeasible-interior-point algorithm whose iterates are confined to the neighborhood* $\mathcal{N}_{-\infty}(\gamma, \beta)$ *defined by (6.2), and that* $\nu_k \downarrow 0$ *for* ν_k *defined by (6.7). Then if* $(\Delta x^k, \Delta\lambda^k, \Delta s^k)$ *is the affine-scaling step defined by (7.2) at the kth iterate, there is a constant* C_5 *such that*

$$\|(\Delta x^k, \Delta s^k)\| \le C_5 \mu_k$$

for all μ_k *sufficiently small.*

Proof. The structure of the proof is the same as in the feasible case. We give part of it and outline the rest, leaving the reader to fill in the last details.

The first part is a generalization of Lemma 7.1; we seek a constant C_4 such that

$$\|\Delta x_{\mathcal{N}}^k\| \le C_4 \mu_k, \qquad \|\Delta s_{\mathcal{B}}^k\| \le C_4 \mu_k. \tag{7.16}$$

These estimates follow from results of Chapter 6. In Theorem 6.8, we showed that there is a constant C_3 such that

$$0 < x_i^k \le \mu_k / C_3, \qquad s_i^k \ge C_3 \gamma$$

for all $i \in \mathcal{N}$. (That particular proof was concerned with the algorithm of Chapter 6, but the logic remains true for the assumptions stated above.) Lemma 6.5 shows that there is a constant C_2 such that

$$\|(D^k)^{-1} \Delta x^k\| \le C_2 \mu_k^{1/2}.$$

As in Lemma 7.1, we note that $D_{ii}^k = \sqrt{x_i^k / s_i^k}$, so we have for all $i \in \mathcal{N}$ that

$$(\Delta x_i^k)^2 = \frac{x_i^k}{s_i^k} \left[(D^k)_{ii}^{-1} \Delta x_i^k \right]^2 \le \frac{x_i^k}{s_i^k} \|(D^k)^{-1} \Delta x^k\|^2 \le \frac{\mu_k}{C_3^2 \gamma} C_2^2 \mu_k.$$

Hence, $\|\Delta x_{\mathcal{N}}^k\| \le C_4 \mu_k$ for an obvious definition of C_4. The bounds on $\|\Delta s_{\mathcal{B}}\|$ follows similarly.

The second part of the proof is a modification of Lemma 7.3 in which the residual terms r_b^k and r_c^k are introduced into the two quadratic programs (7.8) and (7.9). The fact that $\|(r_b^k, r_c^k)\|$ is $O(\mu_k)$ is used to obtain the final estimate. We leave the reader to fill in the details (see the exercises). $\square$

Algorithm PC Is Superlinear

We now return to Algorithm PC from Chapter 5, the path-following algorithm in which affine-scaling (predictor) steps alternate with pure centering (corrector) steps. Each predictor step starts inside an inner neighborhood $\mathcal{N}_2(0.25)$ and chooses α to be the step length to the boundary of an outer neighborhood $\mathcal{N}_2(0.5)$. In Chapter 5, we showed that such steps have length at least $0.4/\sqrt{n}$, so that the method is polynomial-time. Here, we revisit this analysis to show that the predictor step lengths approach 1 on later iterations and therefore that Algorithm PC is superlinearly convergent. This result was proven by Ye et al. [158].

In the key result, we reintroduce the notation $(x(\alpha), \lambda(\alpha), s(\alpha))$ and $\mu(\alpha)$ defined in (5.5).

Lemma 7.6 *Suppose that* $(x, \lambda, s) \in \mathcal{N}_2(0.25)$*, and let* $(\Delta x, \Delta\lambda, \Delta s)$ *be calculated from* (7.1)*. Then if we choose* α *to be the largest value for which* $(x(\alpha), \lambda(\alpha), s(\alpha)) \in \mathcal{N}_2(0.5)$*, we have*

$$\alpha \geq 1 - 4C_5^2\mu, \tag{7.17}$$

where C_5 *is defined from Theorem* 7.4 *with* $\gamma = 0.5$.

Proof. As in the proof of Lemma 5.7, we have

$$\begin{aligned} \|X(\alpha)S(\alpha)e - \mu(\alpha)e\| &\leq (1-\alpha)\|XSe - \mu e\| + \alpha^2\|\Delta X \Delta S e\| \\ &\leq (1-\alpha)(0.25\mu) + \alpha^2 C_5^2 \mu^2, \end{aligned}$$

where we used $(x, \lambda, s) \in \mathcal{N}_2(0.25) \subset \mathcal{N}_{-\infty}(.5)$ to derive the second inequality from Theorem 7.4. Since $\mu(\alpha) = (1-\alpha)\mu$, a sufficient condition for $(x(\alpha), \lambda(\alpha), s(\alpha)) \in \mathcal{N}_2(0.5)$ is that

$$(1-\alpha)(0.25\mu) + \alpha^2 C_5^2 \mu^2 \leq (1-\alpha)(0.5\mu).$$

We seek the largest value of α for which this inequality holds. Rearranging and dividing by $\mu/4$, we obtain

$$(4C_5^2\mu)\alpha^2 + \alpha - 1 \leq 0.$$

By finding the roots of the quadratic, we deduce that the inequality holds for all α in the range

$$\alpha \in \left[0, \frac{\sqrt{1 + 16C_5^2\mu} - 1}{8C_5^2\mu}\right].$$

Elementary manipulation of the upper limit yields

$$\frac{-1+\sqrt{1+16C_5^2\mu}}{8C_5^2\mu} = 1 - \frac{16C_5^2\mu}{\left(\sqrt{1+16C_5^2\mu}+1\right)^2} \geq 1 - 4C_5^2\mu. \tag{7.18}$$

We have shown that $(x(\alpha), \lambda(\alpha), s(\alpha)) \in \mathcal{N}_2(0.5)$ for all $\alpha \in [0, 1-4C_5^2\mu]$, so the proof is complete. □

Fast convergence is an immediate consequence.

Theorem 7.7 *For Algorithm PC, we have*

$$\mu_{k+1} \leq 4C_5^2\mu_k^2, \qquad k = 0, 2, 4, \dots, \tag{7.19a}$$

$$\mu_{k+1} = \mu_k, \qquad k = 1, 3, 5, \dots, \tag{7.19b}$$

where C_5 is defined from Theorem 7.4 with $\gamma = 0.5$.

Proof. For even values of k, we have from Lemma 7.6 that

$$\mu_{k+1} = (1-\alpha_k)\mu_k \leq 4C_5^2\mu_k^2,$$

whereas for odd values of k, the duality measure is unchanged as before. □

By combining the two equations in (7.19), we have that

$$\mu_{k+2} \leq 4C_5^2\mu_k^2, \qquad k = 0, 1, 2, 3, \dots, \tag{7.20}$$

which is referred to as *two-step quadratic* convergence.

Nearly Quadratic Methods

Algorithm PC requires two steps to obtain a quadratic reduction in μ. This performance doesn't quite match the standard performance of Newton's method, which requires just one step to achieve the same effect. We now develop a modification of another algorithm from Chapter 5—Algorithm LPF—that achieves almost quadratic reduction at every step. We will call the modified method *Algorithm LPF+*.

Unlike Algorithm LPF and the other algorithms of Chapters 4, 5, and 6, the new algorithm may need to solve more than one linear system of equations at each iteration. When this need arises, however, the coefficient matrix of each of these systems is the same—only the right-hand side changes. Hence, factorization of the coefficient matrix, which accounts for most of the computational expense, need be performed only once, and each iteration of Algorithm LPF+ is only marginally more expensive than each iteration of the algorithms of Chapters 4, 5, and 6.

Algorithm LPF+ takes two kinds of steps:

- *fast* steps, which are just affine-scaling steps, like the predictor steps in Algorithm PC;
- *safe* steps, which are identical to those taken by the original algorithm LPF.

Recall that in Algorithm LPF, there is a positive lower bound $\sigma_{\min}$ on the value of the centering parameter σ_k. The need for this lower bound becomes clear in the proof of Theorem 5.11: it ensures that we are able to move at least a tiny distance along (x^k, λ^k, s^k) without leaving the neighborhood $\mathcal{N}_{-\infty}(\gamma)$. In particular, (5.22) shows that the minimum value of α_k is proportional to σ_k, where $\sigma_k \geq \sigma_{\min} > 0$. The downside of this argument is that if we set $\sigma_k = 0$—giving an affine-scaling search direction—it may not be possible to move any distance at all along (x^k, λ^k, s^k) without leaving the neighborhood $\mathcal{N}_{-\infty}(\gamma)$.

The way around this difficulty is to reduce the value of γ slightly when we take an affine-scaling step. Reducing γ has the effect of "opening the cone," that is, expanding the neighborhood $\mathcal{N}_{-\infty}(\gamma)$ a little. Because the neighborhood is larger, we can take a nonzero step along the affine-scaling direction before reaching the boundary. There are limits, of course, to how much we can open the cone, since the value of γ must remain strictly positive. If we decrease γ at one step, we leave less room to decrease it later. Fortunately, though, if we take a near-unit step, the affine-scaling direction at the next iteration will be much shorter, so we need decrease γ only by a relatively small amount to achieve a near-unit step length at the next iteration.

The process of opening the cone is illustrated in Figure 7.1. At the kth iterate, the affine-scaling direction points outside the current neighborhood $\mathcal{N}_{-\infty}(\gamma_k)$, but by reducing γ to γ_{k+1} we are able to take a long step. At the $(k+1)$st iterate, the affine-scaling step is shorter, so by decreasing γ just a *little* more (to γ_{k+2}) we are able to take a near-unit step along the $(k+1)$st search direction. Eventually, Algorithm LPF+ is able to take near-unit affine-scaling steps at every iteration, and so it converges superlinearly.

To make the outline of Algorithm LPF+ a little clearer, we combine the process of computing the search direction and choosing α into a single function called **long_step**.

function long_step $((x, \lambda, s), \tilde{\gamma}, \tilde{\sigma})$
solve (1.12) with $\sigma = \tilde{\sigma}$ to obtain $(\Delta x, \Delta \lambda, \Delta s)$;
choose $\tilde{\alpha}$ to be the largest value in $[0, 1]$ for which
    $(x(\tilde{\alpha}), \lambda(\tilde{\alpha}), s(\tilde{\alpha})) \in \mathcal{N}_{-\infty}(\tilde{\gamma})$;
return $\tilde{\alpha}$.

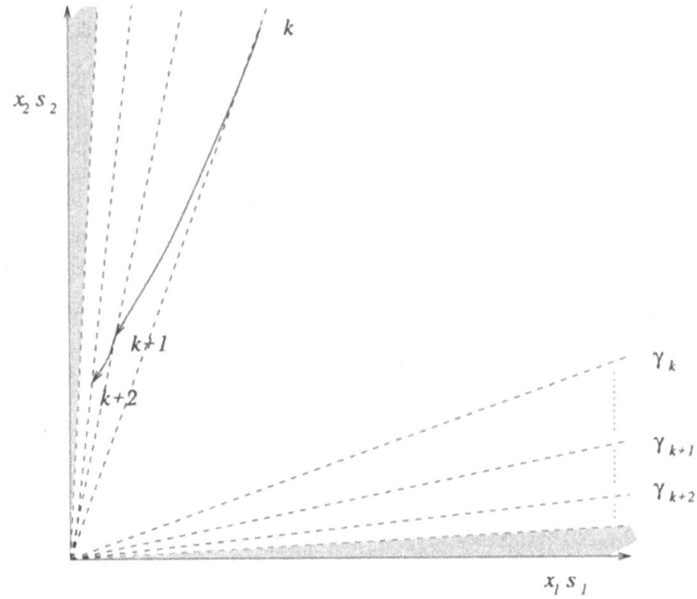

Figure 7.1. Opening the $\mathcal{N}_{-\infty}(\gamma)$ cone to allow nontrivial steps along affine-scaling directions.

Algorithm LPF+ requires a few extra parameters in addition to those used in Algorithm LPF. These parameters, and their allowable ranges, are as follows:

$$0 < \gamma_{\min} < \gamma_{\max} < 1, \quad 0 < \bar{\gamma} \leq 0.5, \quad 0 < \sigma_{\min} < \sigma_{\max} < 1, \quad 0 < \kappa < \bar{\gamma}.$$

Typical values are

$$\begin{aligned} \gamma_{\min} &= 0.001, & \gamma_{\max} &= 0.01, & \bar{\gamma} &= 0.5, \\ \sigma_{\min} &= 0.01, & \sigma_{\max} &= 0.8, & \kappa &= 0.4. \end{aligned}$$

At each iteration, we calculate the affine-scaling direction, decrease γ a little to open the cone, and calculate how far we can move along this direction. If the value of α exceeds $1 - \kappa$, where κ is the parameter defined above, we take the step and label it a *fast step*. Otherwise, we discard this direction, compute a direction of the type used in Algorithm LPF, and take a step *without* opening the cone. This is a *safe step*. The algorithm maintains a counter t_k of the number of fast steps taken prior to iteration k.

Algorithm LPF+

Given $\gamma_{\min}$, $\gamma_{\max}$, $\bar{\gamma}$, $\sigma_{\min}$, $\sigma_{\max}$, and κ,
  and $(x^0, \lambda^0, s^0) \in \mathcal{N}_{-\infty}(\gamma_{\max})$;
$t_0 = 0$; $\gamma_0 \leftarrow \gamma_{\max}$;
for $k = 0, 1, 2, \ldots$
  (* attempt a fast step *)
  $\sigma_k \leftarrow 0$; $\gamma_{k+1} \leftarrow (1-\bar{\gamma})\gamma_{\min} + \bar{\gamma}\gamma_k$;
  $\alpha_k \leftarrow$ **long_step** $((x^k, \lambda^k, s^k), \gamma_{k+1}, \sigma_k)$;
  if $\alpha_k \geq 1 - \kappa$
    (* successful fast step *)
    $(x^{k+1}, \lambda^{k+1}, s^{k+1}) \leftarrow (x^k(\alpha_k), \lambda^k(\alpha_k), s^k(\alpha_k))$;
    $t_{k+1} \leftarrow t_k + 1$;
  else
    (* fast step failed; take a safe step instead *)
    choose $\sigma_k \in [\sigma_{\min}, \sigma_{\max}]$, $\gamma_{k+1} = \gamma_k$;
    $\alpha_k \leftarrow$ **long_step** $((x^k, \lambda^k, s^k), \gamma_{k+1}, \sigma_k)$;
    $(x^{k+1}, \lambda^{k+1}, s^{k+1}) \leftarrow (x^k(\alpha_k), \lambda^k(\alpha_k), s^k(\alpha_k))$;
    $t_{k+1} \leftarrow t_k$;
  end(if)
end(for).

The idea of opening the cone was originally described by Ye and Anstreicher [157] in the context of an algorithm for monotone LCPs that used $\mathcal{N}_2(\cdot)$ neighborhoods. The specification of Algorithm LPF+ and its convergence analysis below follow Wright [146].

Convergence of Algorithm LPF+

The convergence analysis of this method is not difficult, since most of the groundwork has been laid in Chapter 5. The two lemmas below culminate in the main result, Theorem 7.10.

The first result shows that the algorithm reduces the ratio $\mu_k/\bar{\gamma}^{t_k}$ by at least a constant factor at each step, regardless of whether a safe step or fast step is taken.

Lemma 7.8 *For all k, we have*

$$\frac{\mu_{k+1}}{\bar{\gamma}^{t_{k+1}}} \leq \eta \frac{\mu_k}{\bar{\gamma}^{t_k}},$$

where

$$\eta = \max\left(1 - \frac{\delta}{n}, \frac{\kappa}{\bar{\gamma}}\right) < 1,$$

and δ is defined in Theorem 5.11. *Hence, the sequence* $\{\mu_k/\bar{\gamma}^{t_k}\}$ *converges Q-linearly to zero.*

Proof. If a fast step is taken at the kth iteration, we have $\alpha_k \geq 1 - \kappa$. To determine the resulting decrease in μ, we set $\sigma = 0$ in (5.7) and find that

$$\mu_{k+1} = (1 - \alpha_k)\mu_k \leq \kappa\mu_k.$$

The counter t is incremented on a fast step, so that have $t_{k+1} = t_k + 1$. Collecting these observations, we obtain

$$\frac{\mu_{k+1}}{\bar{\gamma}^{t_{k+1}}} \leq \frac{\kappa\mu_k}{\bar{\gamma}\bar{\gamma}^{t_k}} = \left(\frac{\kappa}{\bar{\gamma}}\right)\frac{\mu_k}{\bar{\gamma}^{t_k}}. \tag{7.21}$$

If the fast step is unsuccessful, the algorithm takes a safe step that is exactly the same as a step from the unmodified Algorithm LPF. Hence, from Theorem 5.11, we have $\mu_{k+1} \leq (1 - \delta/n)\mu_k$. The counter t is not incremented, so $t_{k+1} = t_k$. The net result is that

$$\frac{\mu_{k+1}}{\bar{\gamma}^{t_{k+1}}} \leq \left(1 - \frac{\delta}{n}\right)\frac{\mu_k}{\bar{\gamma}^{t_k}}. \tag{7.22}$$

The result comes from combining the two inequalities (7.21) and (7.22). □

The next lemma shows that Algorithm LPF+ takes only fast steps after the ratio $\mu_k/\bar{\gamma}^{t_k}$ falls below a certain threshold.

Lemma 7.9 *Suppose that the condition*

$$\frac{\mu_k}{\bar{\gamma}^{t_k}(\gamma_{\max} - \gamma_{\min})} \leq \frac{\kappa}{2C_5^2} \tag{7.23}$$

holds at some iterate $k = K$, where C_5 is defined from Theorem 7.4 *with $\gamma = \gamma_{\min}$. Then a successful fast step is taken at every iteration $k \geq K$ with step length α_k satisfying*

$$\alpha_k \geq 1 - 2C_5^2 \frac{\mu_k}{\bar{\gamma}^{t_k}(\gamma_{\max} - \gamma_{\min})}. \tag{7.24}$$

Hence, we have

$$\mu_{k+1} \leq \frac{2C_5^2}{\bar{\gamma}^{t_k}(\gamma_{\max} - \gamma_{\min})}\mu_k^2 \qquad \text{for all } k \geq K. \tag{7.25}$$

Proof. Suppose that the condition (7.23) holds for some k. We will show that for all α in the range

$$0 \leq \alpha \leq 1 - 2C_5^2 \frac{\mu_k}{\bar{\gamma}^{t_k}(\gamma_{\max} - \gamma_{\min})}, \tag{7.26}$$

the following condition holds:

$$(x^k(\alpha), \lambda^k(\alpha), s^k(\alpha)) \in \mathcal{N}_{-\infty}(\gamma_{k+1}), \tag{7.27}$$

where γ_{k+1} is the value assigned in the fast branch of Algorithm LPF+. Specifically, we require that

$$\begin{aligned} (x_i^k + \alpha \Delta x_i^k)(s_i^k + \alpha \Delta s_i^k) &\geq \gamma_{k+1}(x^k + \alpha \Delta x^k)^T (s^k + \alpha \Delta s^k)/n \\ &= \gamma_{k+1}(1-\alpha)\mu_k. \end{aligned} \tag{7.28}$$

It follows from this bound that the function **long_step** will choose the actual step to be at least as long as the upper bound, and so (7.24) is true.

Let us see how the counter t_k relates to γ_k for each k. Rearranging the formula for γ_{k+1} in the fast step branch, we obtain

$$\begin{aligned} \gamma_{k+1} - \gamma_{\min} &= \bar{\gamma}(\gamma_k - \gamma_{\min}) && \text{if } k \text{ is a fast step}, \\ \gamma_{k+1} - \gamma_{\min} &= \gamma_k - \gamma_{\min} && \text{if } k \text{ is a safe step}, \end{aligned}$$

so that the gap between γ and $\gamma_{\min}$ is decreased by a factor of $\bar{\gamma}$ at each fast step. Therefore, we have

$$\gamma_{k+1} - \gamma_{\min} = \bar{\gamma}^{t_{k+1}}(\gamma_0 - \gamma_{\min}) = \bar{\gamma}^{t_{k+1}}(\gamma_{\max} - \gamma_{\min}) \tag{7.29}$$

for all $k \geq 0$. If a fast step is taken at step k, we have $t_{k+1} = t_k + 1$, so (7.29) implies that

$$\begin{aligned} \gamma_k - \gamma_{k+1} &= \bar{\gamma}^{t_k}(\gamma_{\max} - \gamma_{\min}) - \bar{\gamma}^{t_{k+1}}(\gamma_{\max} - \gamma_{\min}) \\ &= (1 - \bar{\gamma})\bar{\gamma}^{t_k}(\gamma_{\max} - \gamma_{\min}) \\ &\geq \tfrac{1}{2}\bar{\gamma}^{t_k}(\gamma_{\max} - \gamma_{\min}), \end{aligned} \tag{7.30}$$

where the final inequality comes from $\bar{\gamma} \in (0, .5]$.

Checking the pairwise products along the fast step, we have from the last block row in (7.1) that

$$(x_i^k + \alpha \Delta x_i^k)(s_i^k + \alpha \Delta s_i^k) = x_i^k s_i^k (1 - \alpha) + \alpha^2 \Delta x_i^k \Delta s_i^k. \tag{7.31}$$

Since $(x^k, \lambda^k, s^k) \in \mathcal{N}_{-\infty}(\gamma_k)$, we have for the first term that $x_i^k s_i^k(1-\alpha) \geq \gamma_k \mu_k (1-\alpha)$. Therefore, (7.30) implies that

$$x_i^k s_i^k(1-\alpha) \geq \mu_k(1-\alpha)\left[\gamma_{k+1} + \tfrac{1}{2}\bar{\gamma}^{t_k}(\gamma_{\max} - \gamma_{\min})\right]. \tag{7.32}$$

For the second term in (7.31), we note that $(x^k, \lambda^k, s^k) \in \mathcal{N}_{-\infty}(\gamma_k) \subset \mathcal{N}_{-\infty}(\gamma_{\min})$ and apply Theorem 7.4 to deduce that

$$\alpha^2 \Delta x_i^k \Delta s_i^k \geq -|\Delta x_i^k||\Delta s_i^k| \geq -\|\Delta x^k\|\|\Delta s^k\| \geq -C_5^2 \mu_k^2. \tag{7.33}$$

Simple rearrangement of the condition (7.26) gives

$$-C_5^2 \mu_k^2 \geq -\tfrac{1}{2}\mu_k(1-\alpha)\bar{\gamma}^{t_k}(\gamma_{\max} - \gamma_{\min}). \tag{7.34}$$

We can now substitute (7.32), (7.33), and (7.34) into (7.31) to obtain

$$\begin{aligned}
&(x_i^k + \alpha \Delta x_i^k)(s_i^k + \alpha \Delta s_i^k) \\
&\geq \mu_k(1-\alpha)\left[\gamma_{k+1} + \tfrac{1}{2}\bar{\gamma}^{t_k}(\gamma_{\max} - \gamma_{\min})\right] - \mu_k(1-\alpha)\tfrac{1}{2}\bar{\gamma}^{t_k}(\gamma_{\max} - \gamma_{\min}) \\
&= \mu_k(1-\alpha)\gamma_{k+1}.
\end{aligned}$$

Therefore, for the value of γ_{k+1} assigned in the fast branch and for all α in the range (7.26), we have shown that (7.28) holds, as required. The function **long_step** will therefore select α_k to satisfy (7.24). Since $\mu_{k+1} = (1-\alpha_k)\mu_k$, the bound (7.25) follows immediately. □

Lemma 7.9 captures the actual behavior of Algorithm LPF+. At early iterations, it usually takes safe steps because the affine-scaling direction is too aggressive to make much progress. Near convergence, it takes only fast steps, as predicted by Lemma 7.9. There is sometimes also an intermediate phase in which safe and fast steps are intermingled.

We can figure the rate of convergence from the formula (7.25). This bound indicates superlinear convergence, since

$$\frac{\mu_{k+1}}{\mu_k} \leq \frac{2C_5^2}{(\gamma_{\max} - \gamma_{\min})} \frac{\mu_k}{\bar{\gamma}^{t_k}} \leq \frac{2C_5^2}{(\gamma_{\max} - \gamma_{\min})} \eta^k \mu_0 \to 0, \tag{7.35}$$

where η is defined in Lemma 7.8. It is not quite quadratic convergence, but it is close to quadratic, as we see in the main theorem.

Theorem 7.10 *The sequence $\{\mu_k\}$ converges to zero with Q-order 2, that is,*

$$\liminf_{k \to \infty} \frac{\log \mu_{k+1}}{\log \mu_k} \geq 2. \tag{7.36}$$

Proof. By taking logarithms of both sides in (7.25), we have

$$\log \mu_{k+1} \leq \log \left(2C_5^2/(\gamma_{\max} - \gamma_{\min})\right) - t_k \log \bar{\gamma} + 2 \log \mu_k.$$

Since $\log \mu_k \downarrow -\infty$, we can divide both sides by $\log \mu_k$ for k sufficiently large to obtain

$$\frac{\log \mu_{k+1}}{\log \mu_k} \geq 2 + \frac{\log \left(2C_5^2/(\gamma_{\max} - \gamma_{\min})\right)}{\log \mu_k} - \frac{t_k}{\log \mu_k} \log \bar{\gamma}. \tag{7.37}$$

(The inequality has switched because $\log \mu_k$ is negative for k sufficiently large.) It is clear that the second term on the right-hand side goes to zero as $k \to \infty$. We claim that the third term also goes to zero; that is, $|\log \mu_k|$ grows faster than t_k. Since $0 \leq t_k \leq k$, our claim will hold if we can show that

$$\lim_{k \to \infty} \frac{k}{\log \mu_k} = 0. \tag{7.38}$$

Suppose for contradiction that there is a subsequence $\mathcal{K}$ and a constant $\xi > 0$ such that

$$\frac{k}{\log \mu_k} \leq \frac{1}{\log \xi} \qquad \text{for all } k \in \mathcal{K}.$$

This inequality is equivalent to

$$\xi^k \leq \mu_k \qquad \text{for all } k \in \mathcal{K}.$$

From (7.35), there is an index J such that

$$\mu_{k+1} \leq \frac{\xi}{2} \mu_k \qquad \text{for all } k \geq J.$$

Hence, for all $k \in \mathcal{K}$ with $k \geq J$, we have

$$\xi^k \leq \mu_k \leq \left(\frac{\xi}{2}\right)^{k-J} \mu_J$$

so that

$$2^k \leq \mu_J 2^J / \xi^J.$$

This last expression clearly cannot hold for large k, since the right-hand side is a constant. Hence we have the contradiction that proves (7.38).

Returning to (7.37), we have

$$\liminf_{k \to \infty} \frac{\log \mu_{k+1}}{\log \mu_k} \geq 2,$$

so the rate of convergence has Q-order of at least 2, as claimed. □

Convergence of the Iteration Sequence

In Chapters 5 and 6, we showed that the limit points of the sequences $\{(x^k, s^k)\}$ generated by path-following algorithms can be used to construct strictly complementary primal-dual solutions. When a strictly feasible primal-dual point exists, Theorems 5.14 and 6.9 imply that the sequences $\{(x^k, s^k)\}$ are indeed bounded and therefore that limit points exist.

The analysis of Lemma 5.13 and Theorem 5.14 continues to hold for Algorithm LPF+, since it is a path-following method whose iterates are confined to the neighborhood $\mathcal{N}_{-\infty}(\gamma_{\min})$. In fact, we can prove a stronger result for this algorithm. The sequence $\{(x^k, s^k)\}$ isn't just bounded—it actually converges!

Theorem 7.11 *The sequence $\{(x^k, s^k)\}$ generated by Algorithm LPF+ converges R-superlinearly to a point (x^*, s^*), and there is a vector $\lambda^* \in \mathbb{R}^m$ such that (x^*, λ^*, s^*) is a strictly complementary solution.*

Proof. We first show convergence to a point (x^*, s^*). Let K be the index defined in Lemma 7.9, so that fast steps are taken at all iterates $k \geq K$ and we have $\|(\Delta x^k, \Delta s^k)\| \leq C_5 \mu_k$. We also have $\mu_{k+1} \leq \kappa \mu_k$, since otherwise the fast steps would be rejected by Algorithm LPF+.

Given this information, it is easy to show that the sequence $\{(x^k, s^k)\}$ is Cauchy. For any two indices k_1 and k_2 with $k_2 > k_1 \geq K$, we have

$$\begin{aligned}
\|(x^{k_2}, s^{k_2}) - (x^{k_1}, s^{k_1})\| &= \left\| \sum_{j=k_1}^{k_2-1} \alpha_j (\Delta x^j, \Delta s^j) \right\| \\
&\leq \sum_{j=k_1}^{k_2-1} \|(\Delta x^j, \Delta s^j)\| \\
&\leq C_5 \sum_{j=k_1}^{k_2-1} \mu_j \\
&\leq C_5 \sum_{j=k_1}^{k_2-1} \kappa^{j-k_1} \mu_{k_1} \\
&< \frac{C_5}{1-\kappa} \mu_{k_1}.
\end{aligned}$$

Since $\mu_k \downarrow 0$, we obtain

$$\|(x^{k_2}, s^{k_2}) - (x^{k_1}, s^{k_1})\| \to 0 \qquad \text{as } k_1, k_2 \to \infty,$$

so the sequence $\{(x^k, s^k)\}$ is Cauchy, and therefore it converges to a limit point (x^*, s^*). The analysis of Lemma 5.13 (with $\gamma = \gamma_{\min}$) now shows that (x^*, λ^*, s^*) is a strictly complementary solution for some vector $\lambda^* \in \mathbb{R}^m$.

To show R-superlinear convergence, we obtain by the same argument as above that

$$\|(x^k, s^k) - (x^*, s^*)\| \leq \frac{C_5}{1-\kappa}\mu_k$$

for all $k \geq K$. Since μ_k converges Q-superlinearly to zero, it follows immediately from this bound that $\{(x^k, s^k)\}$ converges R-superlinearly to (x^*, s^*). □

$\epsilon(A, b, c)$ and Finite Termination

A careful review of the analysis in this section reveals that the quantity $\epsilon(A, b, c)$ defined in (3.5) governs the threshold value of μ below which superlinear convergence takes hold. From (7.4), (7.6), and the proof of Theorem 7.4, we can see that

$$C_3 \sim \epsilon(A, b, c), \qquad C_4 \sim \epsilon(A, b, c)^{-1}, \qquad C_5 \sim \epsilon(A, b, c)^{-1}$$

(where the symbol $\sim$ should be read as "varies like"). The analysis of Lemma 7.6 suggests that superlinear convergence becomes noticeable in Algorithm PC when the bound in (7.17) approaches 1. This happens when μ becomes smaller than C_5^{-2} or, to put it another way,

$$\mu \lesssim \epsilon(A, b, c)^2 \tag{7.39}$$

(reading $\lesssim$ as "less than about"). Similarly, for Algorithm LPF+, we see from Lemma 7.9 that the threshold below which fast steps are taken is some fraction of C_5^{-2}, so the criterion (7.39) applies to this case as well.

Once we enter the regime defined by (7.39), the partition of indices into $\mathcal{B}$ and $\mathcal{N}$ can be deduced by inspection of the iterates (x, λ, s). It is fairly clear that the $x_{\mathcal{N}}$ and $s_{\mathcal{B}}$ components are going to zero, whereas the $x_{\mathcal{B}}$ and $s_{\mathcal{N}}$ components are headed toward strictly positive values (see (5.29) and (5.30)). A little thought might lead us to try the following strategy: When we are fairly certain about the contents of $\mathcal{B}$ and $\mathcal{N}$, we

- set $x_{\mathcal{N}}$ and $s_{\mathcal{B}}$ to zero;
- adjust $x_{\mathcal{B}}$, $s_{\mathcal{N}}$, and λ so that the constraints $Ax = b$, $A^T\lambda + s = c$ are satisfied;

- check that the values of $x_{\mathcal{B}}$ and $s_{\mathcal{N}}$ remain strictly positive after this adjustment has been performed.

These steps are the essence of the *finite termination* strategy, which we now examine.

A Finite Termination Strategy

The finite termination strategy proposed by Ye [155] attempts to jump from a path-following iterate to an exact primal-dual solution. The strategy is successful if the iterate is sufficiently advanced—advanced enough for the index sets $\mathcal{B}$ and $\mathcal{N}$ to be well resolved. If not, the strategy will produce a point that violates one or more of the constraints in (2.1), (2.2). No harm is done when the strategy fails, however; we can simply return to the path-following method and take a few more steps before attempting finite termination again.

Because finite termination can be used with any path-following algorithm, even infeasible-interior-point methods such as Algorithm IPF, the following description is not tied to any particular method. It has the same flavor as the analysis of convergence of the iteration sequence in Chapters 5, where we assumed only that each point (x^k, λ^k, s^k) belonged to a neighborhood $\mathcal{N}_{-\infty}(\gamma)$, not that it was generated by any particular algorithm. For simplicity, we describe the finite termination procedure only for the feasible case and leave the extension to infeasible-interior-point methods as an exercise.

Given any point $(x, \lambda, s) \in \mathcal{N}_{-\infty}(\gamma) \subset \mathcal{F}^o$, we can estimate the basic and nonbasic index sets $\mathcal{B}$ and $\mathcal{N}$ as follows:

$$\begin{aligned} \mathcal{B}(x,s) &= \{i \in \{1,2,\ldots,n\} \mid x_i \geq s_i\}, \\ \mathcal{N}(x,s) &= \{1,2,\ldots,n\} \setminus \mathcal{B}(x,s). \end{aligned} \tag{7.40}$$

This estimate makes sense; at any strictly complementary solution we have by definition of $\mathcal{B}$ and $\mathcal{N}$ that

$$0 = s^*_{\mathcal{B}} < x^*_{\mathcal{B}}, \qquad 0 = x^*_{\mathcal{N}} < s^*_{\mathcal{N}}.$$

Therefore, points $(x, \lambda, s) \in \mathcal{F}^o$ that are *close* to a given strictly complementary solution satisfy $\mathcal{B}(x,s) = \mathcal{B}$ and $\mathcal{N}(x,s) = \mathcal{N}$, that is, the estimates (7.40) are exact.

We now specify Procedure FT, the finite termination procedure of Ye [155].

Procedure FT
Given $\gamma \in (0,1)$ and $(x,\lambda,s) \in \mathcal{N}_{-\infty}(\gamma)$:
Find $\mathcal{B}(x,s)$ and $\mathcal{N}(x,s)$ from (7.40);
Solve the following problem:

$$\min_{(x^*,\lambda^*,s^*)} \tfrac{1}{2}\|x^* - x\|^2 + \tfrac{1}{2}\|s^* - s\|^2, \tag{7.41a}$$

$$Ax^* = b, \qquad A^T\lambda^* + s^* = c, \tag{7.41b}$$

$$x_i^* = 0 \text{ for } i \in \mathcal{N}(x,s), \qquad s_i^* = 0 \text{ for } i \in \mathcal{B}(x,s). \tag{7.41c}$$

if $x_{\mathcal{B}}^* > 0$ and $s_{\mathcal{N}}^* > 0$
declare success: (x^*,λ^*,s^*) is a strictly complementary solution;
else
declare failure and return to the path-following algorithm.

The problem (7.41) is an equality-constrained linear least-squares problem, which costs about as much to solve as one iteration of an interior-point algorithm. It can be decoupled trivially (as in Ye [155]) into two separate problems whose unknowns are x^* and (λ^*, s^*), but it is slightly more convenient for the purpose of our analysis to consider the combined form. Any solution of (7.41) satisfies most of the KKT conditions (2.4) for optimality. The constraints (7.41b) imply (2.4a) and (2.4b), whereas (7.41c) and the definition of $\mathcal{B}(x,s)$ and $\mathcal{N}(x,s)$ imply complementarity (2.4c). The only other condition needed for optimality is nonnegativity (2.4d); that is, $(x^*,s^*) \geq 0$. From (7.41c), this condition is satisfied for half of the components of (x^*,s^*). The **if** statement in Procedure FT checks that the condition also is satisfied for the remaining components x_i^* ($i \in \mathcal{B}(x,s)$) and s_i^* ($i \in \mathcal{N}(x,s)$). If this test succeeds, (x^*,λ^*,s^*) is a primal-dual solution.

The following result shows that a successful outcome for Procedure FT is guaranteed when μ is sufficiently small.

Theorem 7.12 *Let $\gamma \in (0,1)$ be given. Then there is a threshold value $\bar{\mu}$ such that for all (x,λ,s) that satisfy*

$$(x,\lambda,s) \in \mathcal{N}_{-\infty}(\gamma) \subset \mathcal{F}^o, \qquad 0 < \mu = x^Ts/n \leq \bar{\mu},$$

we have

(i) *$\mathcal{B}(x,s) = \mathcal{B}$ and $\mathcal{N}(x,s) = \mathcal{N}$—that is, the actual index sets $\mathcal{B}$ and $\mathcal{N}$ are estimated correctly by the procedure* (7.40);

(ii) *the projection procedure* (7.41) *yields a strictly complementary solution* (x^*, λ^*, s^*).

Proof. For (i), we again use the results (5.29) and (5.30) from Lemma 5.13. In particular, we have

$$0 < s_i < \mu/C_3, \qquad x_i \geq C_3\gamma \qquad \text{for all } i \in \mathcal{B}, \tag{7.42}$$

where $C_3 = \epsilon(A, b, c)/n$ as defined in (7.4). By setting

$$\mu \leq \tfrac{1}{2}C_3^2\gamma = \epsilon(A, b, c)^2 \frac{\gamma}{2n^2}, \tag{7.43}$$

we deduce from (7.42) that $0 < s_i < x_i$ for all $i \in \mathcal{B}$. Similarly, we can show that $0 < x_i < s_i$ for all $i \in \mathcal{N}$. It follows from (7.40) that $\mathcal{B}(x, s) = \mathcal{B}$ and $\mathcal{N}(x, s) = \mathcal{N}$.

For (ii), we assume that (7.43) holds, so that $\mathcal{B}(x, s) = \mathcal{B}$ and $\mathcal{N}(x, s) = \mathcal{N}$ in (7.41c). Since (x, λ, s) is a feasible point, we can rewrite the constraints (7.41b) and (7.41c) equivalently as

$$\begin{aligned}
A(x^* - x) &= 0, & & \text{(7.44a)}\\
A^T(\lambda^* - \lambda) + (s^* - s) &= 0, & & \text{(7.44b)}\\
(x_i^* - x_i) &= -x_i \quad \text{for all } i \in \mathcal{N}, & & \text{(7.44c)}\\
(s_i^* - s_i) &= -s_i \quad \text{for all } i \in \mathcal{B}. & & \text{(7.44d)}
\end{aligned}$$

Note that these constraints are consistent: they are satisfied for *all* (x^*, λ^*, s^*) in Ω. Hence, we can apply Hoffman's lemma—Lemma A.3 in Appendix A—to the constraint system (7.44). This lemma implies that there is a constant $\hat{C}$ such that (7.44) has a solution $(\tilde{x} - x, \tilde{\lambda} - \lambda, \tilde{s} - s)$ for which

$$\|(\tilde{x} - x, \tilde{\lambda} - \lambda, \tilde{s} - s)\| \leq \hat{C}\|(x_{\mathcal{N}}, s_{\mathcal{B}})\|.$$

Since the vector $(x_{\mathcal{N}}, s_{\mathcal{B}})$ has n components and since each component is smaller than μ/C_3 (from (5.29)), we have

$$\|(\tilde{x} - x, \tilde{\lambda} - \lambda, \tilde{s} - s)\| \leq \mu\hat{C}\sqrt{n}/C_3. \tag{7.45}$$

If (x^*, λ^*, s^*) is the vector that solves the quadratic program (7.41), the value of the objective function (7.41a) is no larger at (x^*, λ^*, s^*) than at $(\tilde{x}, \tilde{\lambda}, \tilde{s})$. It follows from (7.45) that

$$\begin{aligned}
\|x^* - x\|^2 + \|s^* - s\|^2 &\leq \|\tilde{x} - x\|^2 + \|\tilde{s} - s\|^2\\
&\leq \|(\tilde{x} - x, \tilde{\lambda} - \lambda, \tilde{s} - s)\|^2\\
&\leq \mu^2\hat{C}^2 n/C_3^2, \qquad \text{(7.46)}
\end{aligned}$$

and so

$$|x_i^* - x_i| \leq \mu \hat{C}\sqrt{n}/C_3, \qquad |s_i^* - s_i| \leq \mu \hat{C}\sqrt{n}/C_3, \qquad i = 1, 2, \ldots, n.$$

For $i \in \mathcal{B}$, we have from (5.30) that $x_i \geq C_3\gamma$, and therefore

$$x_i^* \geq x_i - |x_i^* - x_i| \geq C_3\gamma - \mu\hat{C}\sqrt{n}/C_3, \qquad i \in \mathcal{B}.$$

Similarly, we have

$$s_i^* \geq s_i - |s_i^* - s_i| \geq C_3\gamma - \mu\hat{C}\sqrt{n}/C_3, \qquad i \in \mathcal{N}.$$

If the duality measure μ satisfies

$$\mu \leq \frac{C_3^2\gamma}{2\hat{C}\sqrt{n}} = \frac{\epsilon(A,b,c)^2\gamma}{2\hat{C}n^{5/2}}, \tag{7.47}$$

then the right-hand sides of the last two bounds are positive, and we have

$$x_{\mathcal{B}}^* > 0, \qquad s_{\mathcal{N}}^* > 0.$$

Hence, the theorem is true if we set $\bar{\mu}$ to the smaller of the two bounds in (7.43) and (7.47), that is,

$$\bar{\mu} = \frac{\epsilon(A,b,c)^2\gamma}{2n^2} \min\left(1, \frac{1}{\hat{C}n^{1/2}}\right). \quad \square \tag{7.48}$$

We see from the definition of $\bar{\mu}$ in Theorem 7.12 that the finite termination procedure succeeds after μ is decreased to the order of $\epsilon(A,b,c)^2$. As anticipated, this threshold is similar to the threshold below which superlinear convergence starts to take effect; see (7.39). Therefore, finite termination might seem to have a clear advantage, since it jumps straight to a solution without iterating any longer than necessary to identify $\mathcal{B}$ and $\mathcal{N}$. In fact, there is not much to choose between the two strategies. Superlinear algorithms take only two or three extra iterations and give solutions that are accurate enough for most purposes. Moreover, the importance of superlinearity is not confined to the last stages of the algorithm. Even outside the domain in which the asymptotic analysis applies, superlinear algorithms tend to make more rapid progress than nonsuperlinear algorithms. Newton's method itself, when applied to nonlinear equations or unconstrained minimization, is a good example of this phenomenon.

Recovering an Optimal Basis

We have seen that primal-dual algorithms yield strictly complementary solutions, for which $x^*_{\mathcal{B}} > 0$ and $s^*_{\mathcal{N}} > 0$. Geometrically, these solutions lie in the interior of the optimal face of the feasible polygon, unlike the vertex solutions produced by the simplex algorithm which are extreme points of the optimal face. For many purposes, interior solutions are desirable or at least acceptable, but there are also applications where vertex solutions, and their corresponding optimal bases, are more useful. Most prominent among the latter are linear programming relaxations of integer programming problems, where it is best from the viewpoint of the branch-and-bound heuristic to find primal solutions with as many zeros as possible. Vertex solutions are useful too when we are solving a sequence of linear programs in which the data for each problem in the sequence differs only slightly from its predecessor. The optimal basis for the first problem in the sequence can be used to hot-start the simplex algorithm for the remaining problems.

Megiddo [86] proposed a simple two-phase method for recovering a vertex solution from an arbitrary solution $(x^*, \lambda^*, s^*) \in \Omega$, assuming that the partition $\mathcal{B} \cup \mathcal{N}$ is known. Both phases take simplex-like steps, so the term "crossover to simplex" often is used in connection with Megiddo's algorithm and others like it.

The first phase of Megiddo's algorithm deals with the primal problem: Starting with an arbitrary initial basis $\mathcal{M}$ and the primal solution x^*, it adjusts both $\mathcal{M}$ and x^* so that $\mathcal{M}$ becomes an optimal basis while x^* becomes the vertex solution corresponding to $\mathcal{M}$. Each iteration begins by selecting an index j such that $j \in \mathcal{B} \backslash \mathcal{M}$ with $x^*_j > 0$. (If no such j exists, we are done.) We now try to drive x^*_j to zero, adjusting the basic variables $x^*_{\mathcal{M}}$ to keep x^* feasible. If it is not possible to move x^*_j to zero without causing some components of $x^*_{\mathcal{M}}$ to go negative, we drive one of the basic components to zero (say, x^*_i, for $i \in \mathcal{M}$) and replace i by j in the basis $\mathcal{M}$.

The algebraic details of a single iteration are quite similar to an iteration of the simplex method. From the feasibility condition $Ax^* = b$ and the definition (2.19) of B as the $m \times m$ column submatrix of A that corresponds to $\mathcal{M}$, we can write

$$\begin{aligned} Ax^* &= Bx^*_{\mathcal{M}} + A_{\cdot j}x^*_j + \sum_{l \notin \mathcal{M}, l \neq j} A_{\cdot l}x^*_l & (7.49) \\ &= B[x^*_{\mathcal{M}} + \alpha z x^*_j] + (1-\alpha)A_{\cdot j}x^*_j + \sum_{l \notin \mathcal{M}, l \neq j} A_{\cdot l}x^*_l, & (7.50) \end{aligned}$$

where $\alpha \in [0,1]$ and $z = B^{-1}A_{\cdot j}$. We now choose α to be the largest value in $[0,1]$ for which the following condition holds:

$$[x^*_{\mathcal{M}} + \alpha z x^*_j] \geq 0.$$

If $\alpha = 1$, we make the substitutions

$$x^*_{\mathcal{M}} \leftarrow x^*_{\mathcal{M}} + z x^*_j, \qquad x^*_j \leftarrow 0,$$

and leave other components x^*_l and the basis $\mathcal{M}$ unchanged. If $\alpha < 1$, on the other hand, there must be an index i such that

$$[x^*_{\mathcal{M}} + \alpha z x^*_j]_i = 0.$$

We now make the substitutions

$$x^*_{\mathcal{M}} \leftarrow x^*_{\mathcal{M}} + \alpha z x^*_j, \qquad x^*_j \leftarrow (1-\alpha) x^*_j,$$

and finally change the basis $\mathcal{M}$ by removing i and adding j.

The primal phase can be summarized as follows:

Procedure OB-P
Given $x^* \in \Omega_P$, the partition $\mathcal{B} \cup \mathcal{N}$, and an initial basis $\mathcal{M}$:
while there exists $j \in \mathcal{B} \backslash \mathcal{M}$ with $x^*_j > 0$
    either move x^*_j to zero (adjusting $x^*_{\mathcal{M}}$ to keep x^* optimal)
        or pivot j into $\mathcal{M}$;
    adjust x^* and $\mathcal{M}$;
end (while).

A little consideration shows that the number of elements j for which $j \in \mathcal{B} \backslash \mathcal{M}$ with $x^*_j > 0$ drops by at least 1 at every iteration. Hence the total number of iterations is no greater than $|\mathcal{B}|$.

The final basis $\mathcal{M}$ from Procedure OB-P may contain indices $i \in \mathcal{M} \cap \mathcal{N}$ for which $s^*_i > 0$. The second phase of Megiddo's algorithm eliminates these elements by adjusting both $\mathcal{M}$ and the dual solution $(\lambda^*, s^*) \in \Omega_D$. The mechanics of this phase are similar to those of the first phase, so we omit the details and summarize it as follows:

Procedure OB-D
Given $(\lambda^*, s^*) \in \Omega_D$, the partition $\mathcal{B} \cup \mathcal{N}$, and the basis $\mathcal{M}$ from
    Procedure OB-P:
while there exists $j \in \mathcal{N} \cap \mathcal{M}$ with $s^*_j > 0$

either move s_j^* to zero (adjusting s_i^*, $i \in \mathcal{N}\backslash\mathcal{M}$ and λ^* to keep
(s^*, λ^*) optimal) or pivot j out of $\mathcal{M}$;
adjust (λ^*, s^*) and $\mathcal{M}$;
end (while).

This phase requires at most $|\mathcal{N}|$ iterations, so the total number of steps for the two phases combined is at most $|\mathcal{B}| + |\mathcal{N}| = n$, making Megiddo's algorithm a polynomial-time algorithm. In practice, the required number of iterations depends on the amount of primal and dual degeneracy.

Since Megiddo's algorithm must be started at an exact solution, it can be applied only in the wake of a finite termination procedure such as Procedure FT above. More efficient variants combine these two procedures and therefore can be started from any sufficiently advanced interior point iterate (x^k, λ^k, s^k). See Andersen and Ye [7] for a description of one such variant.

More on $\epsilon(A, b, c)$

Unfortunately, $\epsilon(A, b, c)$ is not a continuous function of the problem data (A, b, c). An infinitesimal perturbation to the data can turn a large value of $\epsilon(A, b, c)$ into a tiny value. The properties of $\epsilon(A, b, c)$ are not particularly well understood. It does not, for instance, seem to be related to the conditioning of the problem. The problem can be well conditioned in the sense that large perturbations to the problem data (A, b, c) are needed to make it infeasible or unbounded, while at the same time $\epsilon(A, b, c)$ is tiny.

An example of the noncontinuity of $\epsilon(A, b, c)$ with respect to the data is given by the following parametrized linear program in dual form:

$$\max -\lambda_2 \text{ subject to } \begin{bmatrix} -1 & 0 \\ 1 & 0 \\ \delta & -1 \end{bmatrix} \begin{bmatrix} \lambda_1 \\ \lambda_2 \end{bmatrix} + \begin{bmatrix} s_1 \\ s_2 \\ s_3 \end{bmatrix} = \begin{bmatrix} 0 \\ 1 \\ 0 \end{bmatrix}, \quad s \geq 0, \tag{7.51}$$

where $\delta \geq 0$ is the parameter. The corresponding primal problem is

$$\min x_2 \text{ subject to } \begin{bmatrix} -1 & 1 & \delta \\ 0 & 0 & -1 \end{bmatrix} \begin{bmatrix} x_1 \\ x_2 \\ x_3 \end{bmatrix} = \begin{bmatrix} 0 \\ -1 \end{bmatrix}, \quad x \geq 0. \tag{7.52}$$

By adding the conditions $x_i s_i = \tau$ for $\tau > 0$ and $i = 1, 2, 3$ to the constraints in (7.51) and (7.52), we obtain the central path equations (2.25). These

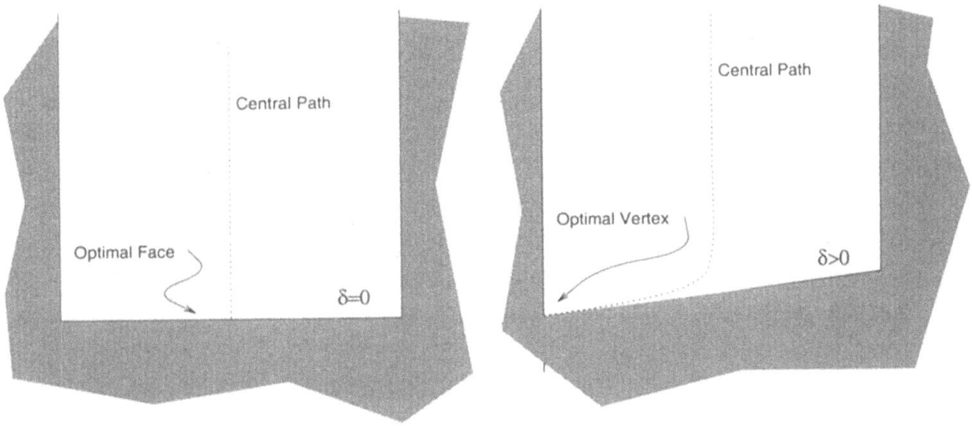

Figure 7.2. Two instances of the problem (7.51) with $\delta = 0$ and δ small and positive and very different values of $\epsilon(A, b, c)$.

equations yield the following solutions for λ:

$$\begin{aligned}\lambda_1 &= \frac{1}{2 + 0.5(\delta/\tau) + 0.125(\delta/\tau)^2}, \\ \lambda_2 &= \frac{\delta}{2 + 0.5(\delta/\tau) + 0.125(\delta/\tau)^2} + \tau.\end{aligned}$$

For the case of $\delta = 0$, the problem has multiple solutions in λ; an entire face of the set of feasible λ values is optimal. Since the unique solutions for x and s are $x^* = (0.5, 0.5, 0)$ and $s^* = (0, 0, 1)$, the definition (3.5) yields the value $\epsilon(A, b, c) = 0.5$. In this case, the x and s component of the central path are

$$s_\tau = (0.5, 0.5, \tau), \qquad x_\tau = (2\tau, 2\tau, 1). \tag{7.53}$$

When τ falls below the threshold $\tau = 0.25$, the partition of indices into $\mathcal{B} = \{3\}$ and $\mathcal{N} = \{1, 2\}$ becomes clearly identifiable. The left-hand diagram in Figure 7.2 illustrates the projection of $\mathcal{C}$ into the space of the λ components. Note that $\mathcal{C}$ approaches the midpoint of the optimal face.

If we now take δ positive and small, the picture changes dramatically. The problem has a unique primal-dual solution given by

$$\lambda^* = (0, 0), \qquad s^* = (0, 1, 0), \qquad x^* = (\delta, 0, 1), \tag{7.54}$$

yielding $\epsilon(A, b, c) = \delta$. The central path is quite smooth for $\tau > \delta$ and seems to be heading for a point close to $\lambda = (0.5, 0)$ as before. However, it takes a sharp turn as τ decreases through δ, and then flattens out again as it approaches the optimum. The situation is illustrated by the right-hand diagram in Figure 7.2.

The problem with $\delta = 0$ is easy to solve, in the sense that superlinear convergence and finite termination properties will take effect even before μ_k becomes particularly small. The problem with δ small and positive is more difficult, since we need to reduce μ_k to a value well below δ before the asymptotic properties take hold. In both cases, the value of $\epsilon(A, b, c)$ is a reliable indicator of the "difficulty" of achieving final convergence to a solution.

This example illustrates that a small perturbation to the data can dramatically change the characteristics of the problem and the value of $\epsilon(A, b, c)$.

Notes and References

The possibility of superlinear convergence is hinted at in Megiddo's original paper [85] on primal-dual methods. After noting that the primal-dual affine-scaling direction is tangent to the central path, he shows in section 6 that the central path becomes almost a straight line near the solution set. (As noted above, the straight-line approach commences when the partition of the indices $\{1, 2, \ldots, n\}$ into $\mathcal{B}$ and $\mathcal{N}$ has resolved itself, that is, when the threshold condition (7.39) is satisfied.)

The connection of primal-dual methods to Newton's method caused a number of later authors to recognize the possibility of superlinear convergence, including Zhang, Tapia, and Dennis [164]. Subsequently, Ye et al. [158] proved that Algorithm PC is two-step quadratic, and Ye and Anstreicher [157] developed their nearly quadratic algorithm for the (more general) LCP.

An infeasible variant of Algorithm LPF+ can be obtained by adding a "fast" branch to Algorithm IPF of Chapter 6. On affine steps, the improvement in duality measure μ is allowed to exceed the improvement in feasibility by a small quantity so that the step lengths for affine-scaling steps approach 1 as above. The convergence rate has Q-order 2, just as in Theorem 7.10. Remarkably, Theorem 7.11 holds for the infeasible variant even when the problem has no strictly feasible points, that is, $\mathcal{F}^o = \emptyset$. See Wright [146] for a complete description of this infeasible and superlinear method in the LCP setting.

When Ye [155] defined Procedure FT, he compared the values of x_i^k and s_i^k for each $i = 1, 2, \ldots, n$ to decide the likely optimal partition $\mathcal{B} \cup \mathcal{N}$.

The resulting criterion (7.40) is but one of many possible techniques for estimating this partition. El-Bakry, Tapia, and Zhang [28] survey a number of these techniques, known as *indicators*, and demonstrate by numerical testing that other indicators may be preferable to (7.40). Of course, it is easy to replace (7.40) by these other indicators in Procedure FT.

Exercises

1. The result of Lemma 7.1 holds not just for $\sigma = 0$ but for all $\sigma \in [0, 1]$ in (1.12). How can the proof be modified to demonstrate this more general result?

2. Prove uniqueness of the $\Delta s_{\mathcal{N}}$ component in Lemma 7.3(ii) in the case in which A does not have full row rank. (Hint: Use the fact that $QA = \left[\begin{smallmatrix}\bar{A}\\0\end{smallmatrix}\right]$, where Q is some orthogonal matrix and $\bar{A}$ has full row rank.)

3. State and prove an infeasible-interior-point version of Lemma 7.3, where $(\Delta x^k, \Delta\lambda^k, \Delta s^k)$ is derived from (7.2) instead of (7.1). (Hint: Find how the systems (7.10) and (7.12) must be changed to accommodate the residual terms r_b^k and r_c^k, and work backward to decide where to put these terms in the formulations (7.8) and (7.9).)

4. Verify (7.18).

5. Show that the convergence of $\{\mu_k\}$ indicated in Theorem 7.7 is *not* Q-superlinear but rather R-superlinear.

6. State and prove an infeasible variant of Theorem 7.12. That is, given $\gamma \in (0, 1)$ and $\beta > 0$, find a threshold value $\bar{\mu}$ such that this result holds for all $(x, \lambda, s) \in \mathcal{N}_{-\infty}(\gamma, \beta)$ with $\mu \le \bar{\mu}$. (Hint: Use the definition of $\mathcal{N}_{-\infty}(\gamma, \beta)$ from (6.2) together with Theorem 6.8.)

7. Suppose that A does not have full rank. Show that if $(x, \lambda, s) \in \mathcal{F}$, we have

$$\text{dist}_\Omega(x, \lambda + v, s) = \text{dist}_\Omega(x, \lambda, s) \qquad \text{for all } v \in \text{Null}(A^T).$$

Chapter 8

Extensions

In this book, our focus is squarely on the linear programming problem. The algorithms, theory, and practical details are described in terms of this problem, because it is (for now) the most important and familiar application of the interior-point approach. We could, however, have rewritten all the exposition in terms of a more general class of problems known as *monotone linear complementarity problems*, which are usually referred to by the acronym LCPs. (The "monotone" is understood.) It is easy to reexpress in terms of LCP all the algorithms we have described for linear programming, and the theoretical properties also extend in a straightforward way. In fact, much of the literature in primal-dual algorithms uses the LCP setting, or one of its equivalent variants, such as mixed or horizontal LCPs. When reading this literature, one must bear in mind that just a few minor notational changes are needed to convert an interior-point algorithm for LCP into a linear programming algorithm.

In this chapter we describe the LCP and a few of its variants and show how the algorithms of earlier chapters can be adapted to these problems. We sketch extensions of primal-dual algorithms to convex QP and to more general problems, including semidefinite programming, convex programming, and monotone nonlinear complementarity problems.

The Monotone LCP

The monotone LCP is to find vectors x and s in $\mathbb{R}^n$ that satisfy the following conditions:

$$s = Mx + q, \qquad (x, s) \geq 0, \qquad x^T s = 0, \tag{8.1}$$

where M is an $n \times n$ positive semidefinite matrix and q is a vector in $\mathbb{R}^n$. The conditions $x^T s = 0$ and $(x, s) \geq 0$ together imply that $x_i s_i = 0$ for each $i = 1, 2, \ldots, n$ or, equivalently,

$$x_i = 0 \text{ or } s_i = 0 \text{ for each } i = 1, 2, \ldots, n.$$

As stated, the problem (8.1) is not an optimization problem because we are not actually minimizing anything. It does, however, have constraints ($s = Mx + q$), nonnegativity conditions ($x \geq 0$ and $s \geq 0$), and a complementarity condition ($x^T s = 0$), so it has a lot in common with the KKT conditions (1.4) for linear programming. As in (1.5), we can rewrite (8.1) as the constrained nonlinear system

$$F(x, s) = \begin{bmatrix} Mx + q - s \\ XSe \end{bmatrix} = 0, \qquad (x, s) \geq 0,$$

where $X = \text{diag}(x_1, x_2, \ldots, x_n)$ and $S = \text{diag}(s_1, s_2, \ldots, s_n)$. As in Chapter 1, we can view primal-dual algorithms as modified Newton methods applied to this system. Framework PD of Chapter 1 can be restated in terms of the LCP as follows.

Framework PD (LCP)
Given (x^0, s^0) with $s^0 = Mx^0 + q$, $(x^0, s^0) > 0$;
for $k = 0, 1, 2, \ldots$
solve

$$\begin{bmatrix} M & -I \\ S^k & X^k \end{bmatrix} \begin{bmatrix} \Delta x^k \\ \Delta s^k \end{bmatrix} = \begin{bmatrix} 0 \\ -X^k S^k e + \sigma_k \mu_k e \end{bmatrix}, \tag{8.2}$$

where $\sigma_k \in [0, 1]$ and $\mu_k = (x^k)^T s^k / n$;

set

$$(x^{k+1}, s^{k+1}) \leftarrow (x^k, s^k) + \alpha_k (\Delta x^k, \Delta s^k), \tag{8.3}$$

choosing α_k so that $(x^{k+1}, s^{k+1}) > 0$.

end (for).

Infeasible-interior-point algorithms also extend to the LCP setting. The only changes needed to Framework PD are that we allow $s^0 \neq Mx^0 + q$ and we replace 0 on the right-hand side of (8.2) by the residual $-r^k = s^k - Mx^k - q$.

The other entities that were used to describe the linear programming algorithms also generalize in fairly obvious ways. For instance, the feasible and strictly feasible regions and are defined as follows (cf. (1.7)):

$$\begin{aligned}\mathcal{F} &= \{(x,s)\,|\,s = Mx+q, (x,s)\geq 0\},\\ \mathcal{F}^o &= \{(x,s)\,|\,s = Mx+q, (x,s) > 0\}.\end{aligned}$$

The central path $\mathcal{C}$ is the set of points (x_τ, s_τ) that satisfy the conditions

$$\begin{aligned}Mx+q &= s,\\ x_i s_i &= \tau, \qquad i = 1,2,\ldots,n,\\ (x,s) &> 0\end{aligned}$$

for some $\tau > 0$. The neighborhoods $\mathcal{N}_2(\theta)$ and $\mathcal{N}_{-\infty}(\gamma)$ are defined analogously to (1.15) and (1.16).

We use Algorithm PC from Chapter 5 to illustrate how linear programming algorithms can be restated in terms of the LCP. Extensions of the other algorithms are left as exercises.

Algorithm PC (for LCP)
Given $(x^0, s^0) \in \mathcal{N}_2(0.25)$;
for $k = 0, 1, 2, \ldots$
  if k is even
    solve (8.2) with $\sigma_k = 0$ to obtain $(\Delta x^k, \Delta s^k)$;
    choose α_k as the largest value of α in $[0,1]$ such that

$$(x^k(\alpha), s^k(\alpha)) \in \mathcal{N}_2(0.5);$$

    set $(x^{k+1}, s^{k+1}) = (x^k, s^k) + \alpha_k(\Delta x^k, \Delta s^k)$;
  else
    solve (8.2) with $\sigma_k = 1$ to obtain $(\Delta x^k, \Delta s^k)$;
    set $(x^{k+1}, s^{k+1}) = (x^k, s^k) + (\Delta x^k, \Delta s^k)$;
  end (if)
end (for).

The global convergence/complexity result for this algorithm (and for extensions of the other algorithms from Chapters 4, 5, and 6) is almost identical to the corresponding result for linear programming (cf. Theorem 5.9).

Theorem 8.1 *Given $\epsilon > 0$, suppose that the starting point $(x^0, s^0) \in \mathcal{N}_2(0.25)$ in Algorithm PC has*

$$\mu_0 \leq \frac{C}{\epsilon^\kappa}$$

for some positive constants C and κ. Then there is an index K with $K = O(\sqrt{n}|\log \epsilon|)$ such that

$$\mu_k \leq \epsilon \qquad \textit{for all } k \geq K.$$

In many ways, LCP provides a more convenient framework for describing and analyzing primal-dual algorithms than does linear programming. LCP problems are fully defined by just two objects (M and q) instead of three; they involve a single equality $s = Mx + q$ instead of the pair of equalities (2.4a), (2.4b), and there are just two vectors of unknowns (x and s) instead of three.

We now describe two variants of the LCP—the mixed LCP and horizontal LCP—that appear to be more general than the standard form (8.1) but are, in fact, equivalent to it. Their usefulness lies in the fact that it is more convenient to state linear and quadratic programs as mixed or horizontal LCPs than to formulate them as standard LCPs.

The equivalence of LCP variants is not merely theoretical or formal. If we apply a given primal-dual algorithm to two different LCP reformulations of the same problem, the kth primal-dual iterate for one formulation can be transformed into the kth primal-dual iterate of the other, provided the starting points are equivalent.

Mixed and Horizontal LCP

In the *mixed* LCP (mLCP), we seek vectors x, s in $\mathbb{R}^n$ and $z \in \mathbb{R}^m$ such that

$$\begin{bmatrix} s \\ 0 \end{bmatrix} = \begin{bmatrix} M_{11} & M_{12} \\ M_{21} & M_{22} \end{bmatrix} \begin{bmatrix} x \\ z \end{bmatrix} + \begin{bmatrix} q_1 \\ q_2 \end{bmatrix}, \quad (x, s) \geq 0, \quad x^T s = 0, \quad (8.4)$$

where the matrix M defined by

$$M = \begin{bmatrix} M_{11} & M_{12} \\ M_{21} & M_{22} \end{bmatrix}$$

is $(m + n) \times (m + n)$ and positive semidefinite. Setting $m = 0$, $M_{11} = M$, and $q_1 = q$, we see that the standard LCP (8.1) is a special case of mLCP. We indicate this relationship by LCP $\subset$ mLCP.

The KKT conditions for linear programming (1.4) constitute an mLCP. We see this by restating them as

$$\begin{bmatrix} s \\ 0 \end{bmatrix} = \begin{bmatrix} 0 & -A^T \\ A & 0 \end{bmatrix} \begin{bmatrix} x \\ \lambda \end{bmatrix} + \begin{bmatrix} c \\ -b \end{bmatrix}, \quad (x,s) \geq 0, \quad x^T s = 0, \tag{8.5}$$

and checking that the coefficient matrix is positive semidefinite (see the exercises).

The *horizontal* LCP (hLCP) is to find x and s in $\mathbb{R}^n$ such that

$$Qs + Rx = q, \qquad (x,s) \geq 0, \qquad x^T s = 0, \tag{8.6}$$

where Q and R are both $n \times n$ matrices with the property that

$$Qs + Rx = 0 \Rightarrow x^T s \geq 0. \tag{8.7}$$

We can see that LCP $\subset$ hLCP by making the identifications $Q = I$ and $R = -M$.

It is trivial to design primal-dual algorithms for each of these variants. Given a current iterate (x, z, s) with $(x, s) > 0$, the basic search direction for mLCP is calculated from

$$\begin{bmatrix} M_{11} & M_{12} & -I \\ M_{21} & M_{22} & 0 \\ S & 0 & X \end{bmatrix} \begin{bmatrix} \Delta x \\ \Delta z \\ \Delta s \end{bmatrix} = \begin{bmatrix} -r_1 \\ -r_2 \\ -XSe + \sigma\mu e \end{bmatrix}, \tag{8.8}$$

where

$$r_1 = M_{11}x + M_{12}z + q_1 - s, \qquad r_2 = M_{21}x + M_{22}z + q_2, \qquad \mu = x^T s/n.$$

(Compare (8.8) with (1.20) for the case of linear programming and (8.2) for standard LCP.) For hLCP the step equation is

$$\begin{bmatrix} R & Q \\ S & X \end{bmatrix} \begin{bmatrix} \Delta x \\ \Delta s \end{bmatrix} = \begin{bmatrix} q - Qs - Rx \\ -XSe + \sigma\mu e \end{bmatrix}. \tag{8.9}$$

LCP is not just a special case of mLCP and hLCP but is actually *equivalent* to these two formulations; that is,

$$\text{mLCP} = \text{hLCP} = \text{LCP}. \tag{8.10}$$

The missing links that we need to complete the proof of this result are given in the following two theorems.

Theorem 8.2 *(hLCP $\subset$ LCP) Any hLCP can be reformulated as an LCP. Primal-dual algorithms applied to the two formulations generate sequences of iterates $\{(x^k, s^k)\}$ that are identical to within a permutation of the variables.*

Theorem 8.3 *(mLCP $\subset$ hLCP) Any mLCP can be reformulated as an hLCP. Primal-dual algorithms applied to the two formulations generate sequences of iterates in which the x^k and s^k components are identical.*

The proof of Theorem 8.2 can be found in Bonnans and Gonzaga [17]. Theorem 8.3 is derived from a number of sources, including Anitescu, Lesaja, and Potra [10] and Wright [146]. These proofs make use of deep results about maximal monotone operators, although they can also be restated in the language of basic linear algebra. We omit them from our present discussion.

Strict Complementarity and LCP

All linear programs in which the primal-dual feasible set $\mathcal{F}$ is nonempty have strictly complementary solutions (see Theorem 2.4 of Chapter 2). This property was important in a number of theorems in Chapters 5, 6, and 7, particularly in results about superlinear convergence and convergence of the iteration sequence.

It is trivial to extend the definition of strict complementarity in Theorem 2.4 to the case of LCP, hLCP, and mLCP. A solution (x^*, s^*) of (8.1) or (8.6) is strictly complementary if $x^* + s^* > 0$, whereas a solution (x^*, z^*, s^*) of the mLCP (8.4) is strictly complementary if $x^* + s^* > 0$.

In the case of LCP, we cannot take for granted the existence of a strictly complementary solution. Some LCPs have solutions but no strictly complementary solutions. A simple example is given by

$$s = x, \qquad s^T x = 0, \qquad (s, x) \geq 0,$$

whose unique solution $s^* = x^* = 0$ is not strictly complementary. Some LCPs have solutions but no strictly complementary solutions are known as *degenerate LCPs*. Algorithmic properties that rely on strict complementarity (most notably, superlinear convergence) fail to hold for degenerate LCPs, although the results on global convergence and polynomial complexity still apply.

Convex QP

Convex QP problems can, like linear programs, be formulated in a number of different ways. The essential features are that the objective function is a convex quadratic and the constraints are linear. One formulation is a straightforward extension of the standard form (2.1) of linear programming:

$$\min \tfrac{1}{2}x^T Qx + c^T x \qquad \text{subject to } Ax = b,\ x \geq 0, \tag{8.11}$$

where Q is a symmetric positive semidefinite $n \times n$ matrix and A, b, and c are as before. The dual for (8.11) is

$$\max -\tfrac{1}{2}x^T Qx + b^T \lambda \text{ subject to } -Qx + A^T \lambda + s = c,\ s \geq 0, \tag{8.12}$$

and the KKT conditions can be conveniently stated in the form of an mLCP as follows:

$$\begin{bmatrix} s \\ 0 \end{bmatrix} = \begin{bmatrix} Q & -A^T \\ A & 0 \end{bmatrix} \begin{bmatrix} x \\ \lambda \end{bmatrix} + \begin{bmatrix} c \\ -b \end{bmatrix}, \quad (x, s) \geq 0, \quad x^T s = 0. \tag{8.13}$$

Introducing slack variables and by splitting free variables into positive and negative parts, we can write *any* convex QP in the form (8.11) and then convert it to an mLCP by using the transformation (8.11) $\to$ (8.13). It is usually simpler, however, to go directly from the KKT conditions for a convex QP to the mLCP form by applying Theorem A.1. We work through some examples of this process in the exercises.

Since convex QP $\subset$ mLCP, we have from Theorems 8.2 and 8.3 that convex QP $\subset$ LCP. The following theorem shows a relationship in the other direction: Any LCP can be expressed as a convex QP. The proof appears at the end of this chapter.

Theorem 8.4 *Given the* LCP (8.1), *define the following convex QP problem:*

$$\min_x x^T(q + Mx) \quad \textit{subject to } q + Mx \geq 0,\ x \geq 0. \tag{8.14}$$

Then x^* *solves* (8.14) *if and only if* $(x^*, Mx^* + q)$ *is a solution of* (8.1).

Unlike the case of hLCP and mLCP, there is no relationship between the primal-dual iterates for the original LCP (8.1) and the primal-dual iterates for the convex QP problem (8.14). Indeed, when we convert (8.14) back into an LCP through its KKT conditions, we obtain a somewhat different LCP from the one that we started with (8.1). In other words, the standard techniques for converting an LCP into a convex QP problem and vice versa do not commute.

Despite the close relationships between the variants of LCP and convex QP, it is worth distinguishing between them because most real-world problems fit more naturally into one of the formulations than the others. On the other hand, it would be wasteful to write a complete piece of software for each problem class, because their close relationships would lead to much duplicated coding effort. The best general-purpose code would probably be one for mLCP, since linear programming (8.5), convex QP (8.14), and standard LCP can all be posed as mLCP without any transformation of the data. If the code is equipped with a variety of linear algebra modules that exploit the structure in each of these problem classes, it loses little if anything in efficiency over individual codes for each of the formulations. For instance, in the case of linear programming, the general step equation obtained by substituting the data of (8.5) into (8.8) leads to the system (1.12). As we see in Chapter 11, this system can be solved efficiently by applying block elimination together with sparse matrix factorization techniques.

Convex Programming

The general convex programming problem is

$$\min_z \phi(z) \qquad \text{subject to } g_i(z) \le 0,\ i = 1, 2, \ldots, m, \tag{8.15}$$

where ϕ and g_i, $i = 1, 2, \ldots, m$ are smooth convex functions. Convexity of the g_i implies convexity of the feasible set defined by

$$\{z \mid g(z) \le 0\}, \qquad \text{where } g(z) = (g_1(z), \ldots, g_m(z))^T;$$

see the exercises. The Jacobian G of the function $g : \mathbb{R}^n \to \mathbb{R}^m$ is defined as

$$G(z) = \left[\frac{\partial g_i}{\partial z_j}\right]_{i=1,\ldots,m,\ j=1,\ldots,n}.$$

The KKT conditions for (8.15) are obtained from a slight extension of the conditions (A.3) in Theorem A.1:

$$\begin{aligned} \nabla\phi(z) + G(z)^T\lambda &= 0, && (8.16a)\\ g(z) + y &= 0, && (8.16b)\\ \lambda_i y_i &= 0, && (8.16c)\\ (\lambda, y) &\ge 0. && (8.16d) \end{aligned}$$

Here, z is the variable vector from (8.15), y is a vector of slack variables for the constraints, and λ is the Lagrange multiplier vector. Writing the Lagrangian function for (8.15) as

$$\mathcal{L}(z, \lambda) = \phi(z) + \lambda^T g(z), \tag{8.17}$$

we can restate the first condition (8.16a) in a succinct manner as

$$\nabla_z \mathcal{L}(z, \lambda) = 0.$$

As with linear programming, the conditions (8.16) can be used to characterize solutions of (8.15), but only if an additional assumption known as a *constraint qualification* is satisfied at the point in question. Essentially, this qualification ensures that the linearized model of the constraint set in a neighborhood of z captures the local behavior of the actual constraint set. Typical constraint qualifications include linearity of the constraints (as in Theorem A.1); linear independence of the active constraint gradients (that is, the submatrix of $G(z)$ containing columns i for which $g_i(z) = 0$); and the Slater condition, which is that there exists a vector z for which $g_i(z) < 0$, all $i = 1, 2, \ldots, m$. See Mangasarian [81] or Fletcher [30, Chapter 9] for more information on constraint qualifications.

Most algorithms for convex programming search for a point that satisfies the KKT conditions (8.16), in the hope that the constraint qualification will hold at this point. It is not difficult to generalize the primal-dual algorithms discussed earlier to this more general setting. Typical iterates (z^k, λ^k, y^k) satisfy the inequality (8.16d) strictly (that is, $(\lambda^k, y^k) > 0$). The typical primal-dual step is, as usual, a modified Newton step for the equality conditions (8.16a)–(8.16c); that is,

$$\begin{bmatrix} \nabla_{zz}\mathcal{L}(z, \lambda) & G(z)^T & 0 \\ G(z) & 0 & I \\ 0 & Y & \Lambda \end{bmatrix} \begin{bmatrix} \Delta z \\ \Delta \lambda \\ \Delta y \end{bmatrix} = \begin{bmatrix} -(\nabla\phi(z) + G(z)^T\lambda) \\ -(g(z) + y) \\ -\Lambda Y e + \sigma\mu e \end{bmatrix}. \tag{8.18}$$

The duality measure μ is defined in this context as

$$\mu = \lambda^T y / m. \tag{8.19}$$

One major difference between linear programming and general convex programming is that line searches along $(\Delta z, \Delta\lambda, \Delta y)$ do not preserve feasibility. For instance, we usually have $g(z + \alpha\Delta z) \neq 0$ even if $g(z) = 0$.

The best way around this difficulty is to use the infeasible-interior-point framework. We simply redefine the neighborhood $\mathcal{N}_{-\infty}(\gamma, \beta)$ from (6.2) as

$$\mathcal{N}_{-\infty}(\gamma,\beta) = \{(z,\lambda,y) \mid \|\nabla_z \mathcal{L}(z,\lambda)\| \le \beta\mu, \quad \|g(z)+y\| \le \beta\mu, \quad (\lambda,y) \ge 0, \quad \lambda_i y_i \ge \gamma\mu, \; i=1,2,\ldots,m\} \tag{8.20}$$

(where $\gamma \in (0,1)$ and $\beta > 0$) and require all iterates to stay within $\mathcal{N}_{-\infty}(\gamma, \beta)$ while reducing μ to zero. The following algorithm, adapted from Ralph and Wright [111], is a natural extension of Algorithm IPF from Chapter 6.

Algorithm IPF (for convex programming)
Given γ, β, $\sigma_{\min}$, $\sigma_{\max}$, χ with $\gamma \in (0,1)$, $\beta > 0$, $0 < \sigma_{\min} < \sigma_{\max} < 1$, $\chi \in (0,1)$, and $(z^0, \lambda^0, y^0) \in \mathcal{N}_{-\infty}(\gamma, \beta)$
for $k = 0, 1, 2, \ldots$
choose $\sigma_k \in [\sigma_{\min}, \sigma_{\max}]$ and solve (8.18) for $(\Delta z, \Delta\lambda, \Delta y)$;
choose α_k as the first element in the sequence $\{1, \chi, \chi^2, \chi^3, \ldots\}$ such that

$$(z^k(\alpha), \lambda^k(\alpha), y^k(\alpha)) \in \mathcal{N}_{-\infty}(\gamma, \beta), \tag{8.21}$$

and

$$\mu_k(\alpha) \le (1 - .01\alpha)\mu_k; \tag{8.22}$$

set $(z^{k+1}, \lambda^{k+1}, y^{k+1}) = (z^k(\alpha_k), \lambda^k(\alpha_k), y^k(\alpha_k))$;
end (for).

It can be shown that all limit points of this algorithm are solutions of (8.15); see [111, section 3] for details.

Special classes of convex programming problems have come under intense scrutiny in the interior-point literature during the past few years. Most activity has been in the area of purely primal methods, which are based on the primal log barrier function and are closely related to the SUMT method of Fiacco and McCormick [29]. Much of the literature is devoted to proving complexity theorems for algorithms for (8.15) (rather than just the simple global convergence result of the previous paragraph), but these theorems require more than mere smoothness of the functions ϕ and g. The barrier function for the constraint set is usually required to satisfy a *self-concordancy condition*, in which its third derivative along any direction is assumed to be bounded in terms of its second derivative at all strictly feasible points (see Nesterov and Nemirovskii [101, page 3]). Fortunately, this class includes many interesting special cases, including semidefinite programming, as we see below. Besides the book of Nesterov and Nemirovskii, information on the self-concordancy condition can be found in the review paper of Todd [128, section 8], the thesis of den Hertog [25], and the papers by Jarre and Saunders [55] and Güler [48].

Nesterov and Todd [102] consider a particular class of constraint sets and their associated self-concordant barrier functions, which they describe as "self-scaled." They describe primal-dual algorithms that use these functions both to generate the search directions and to measure proximity to the central path. The self-scaled class includes such interesting special cases as linear programming, quadratically constrained convex QP, and semidefinite programming (see below) but does not extend to all smooth problems of the form (8.15).

Monotone Nonlinear Complementarity and Variational Inequalities

The monotone LCP (8.1) can be extended into the nonlinear realm by replacing the linear term $Mx + q$ by a monotone nonlinear function. We then have

$$s = f(x), \qquad (x, s) \geq 0, \qquad x^T s = 0, \tag{8.23}$$

where $f : \mathbb{R}^n \to \mathbb{R}^n$ satisfies the monotonicity condition

$$(f(x_1) - f(x_2))^T (x_1 - x_2) \geq 0$$

for all x_1 and x_2 in a neighborhood of the positive orthant $\mathbb{R}^n_+$. The conditions (8.23) themselves form a constrained nonlinear system, and Algorithm IPF (for convex programming) can be adapted for this problem in a straightforward way. The typical primal-dual step is defined by

$$\begin{bmatrix} \nabla f(x) & -I \\ S & X \end{bmatrix} \begin{bmatrix} \Delta x \\ \Delta s \end{bmatrix} = \begin{bmatrix} s - f(x) \\ -XSe + \mu e \end{bmatrix}, \tag{8.24}$$

where $\mu = x^T s/n$. A slightly different approach is described by Kojima, Noma, and Yoshise [69], who "bend" the s component of the search direction so that the residual $s - f(x)$ decreases linearly with step length α.

The monotone variational inequality problem is defined by a monotone mapping $\Phi : \mathbb{R}^n \to \mathbb{R}^n$ and a closed convex set $\mathcal{G} \subset \mathbb{R}^n$. The problem is to find a vector z^* such that

$$z^* \in \mathcal{G}, \qquad (z - z^*)^T \Phi(z^*) \geq 0 \quad \text{for all } z \in \mathcal{G}. \tag{8.25}$$

When the set $\mathcal{G}$ is defined by convex inequalities, that is,

$$\mathcal{G} = \{z \mid g_i(z) \le 0,\ i = 1, 2, \ldots, m\}$$

for convex g_i, we can reformulate the problem (8.25) as a mixed version of the nonlinear complementarity problem (8.23). Under suitable conditions on g, the vector z^* solves (8.25) if and only if there is a vector $\lambda^* \in \mathbb{R}^m$ such that the following conditions hold for $(z, \lambda) = (z^*, \lambda^*)$:

$$\Phi(z) + G(z)^T \lambda = 0, \quad \lambda \ge 0, \quad g(z) \le 0, \quad \lambda^T g(z) = 0.$$

If we make the substitution $y = -g(z)$, we obtain a constrained nonlinear system similar to (8.16), and the convex programming variant of Algorithm IPF can be used to solve it.

Semidefinite Programming

Semidefinite programming (SDP) has been responsible for the latest boomlet in interior-point research. It is an extension of linear programming in which symmetric matrices as well as real vectors are included among the variables, and positive semidefiniteness conditions on the matrix variables are included in the constraints. Interior-point algorithms are well suited for solving SDP, which has a multitude of interesting applications to physical problems, to control problems, and to other areas of mathematical programming. Extension of interior-point algorithms for linear programming to the SDP setting is not necessarily straightforward, however, and the analysis becomes more complicated.

The key to the applicability of interior-point methods to SDP was supplied by Nesterov and Nemirovskii [101, page 245], who showed that a logarithmic barrier function for SDP problems satisfies the self-concordancy condition. Hence, it is possible to devise polynomial interior-point algorithms for this class of problems.

To define the standard form of SDP and its dual, we need a little notation. Let $\mathcal{S}^n$ be the set of real symmetric $n \times n$ matrices, and define an inner product on $\mathcal{S}^n$ by

$$G \bullet H = \operatorname{trace} GH = \sum_{i,j} G_{ij} H_{ij}.$$

If $G \in \mathcal{S}^n$, we use $G \ge 0$ to denote positive semidefiniteness and $G > 0$ to denote positive definiteness.

Using this notation, we define the SDP as

$$\min_{X \in \mathcal{S}^n} C \bullet X \qquad \text{subject to} \quad A(X) = b, \;\; X \succeq 0, \tag{8.26}$$

where

$$A(X) = [A_1 \bullet X, \ldots, A_m \bullet X]^T,$$

with $C \in \mathcal{S}^n$, $A_k \in \mathcal{S}^n$ for all k, and $b \in \mathbb{R}^m$. The dual of (8.26) is

$$\max_{y \in \mathbb{R}^m} b^T y \qquad \text{subject to} \quad A^T(y) + Z = C, \;\; Z \succeq 0, \tag{8.27}$$

where $A^T(y) = \sum_{k=1}^m y_k A_k$ and $Z \in \mathcal{S}^n$ is the dual slack matrix. For any feasible primal-dual triple (X, y, Z), it is easy to check that the duality gap $C \bullet X - b^T y$ is equal to $X \bullet Z$. Accordingly, we define the duality measure for SDP as

$$\mu = \frac{1}{n} X \bullet Z. \tag{8.28}$$

A set of KKT conditions can be used to identify primal and dual solutions when a constraint qualification is satisfied. The solution sets for both the primal problem (8.26) and dual problem (8.27) are nonempty if there exist strictly feasible points for both problems (that is, positive definite matrices X and Z in $\mathcal{S}^n$ and a vector $y \in \mathbb{R}^m$ such that $A(X) = b$ and $A^T(y)+Z = C$); see Vandenberghe and Boyd [135, Theorem 3.1]. The duality measure defined in (8.28) is zero at all such solutions, and the KKT conditions that identify these solutions can be stated analogously to (1.4) as follows:

$$A^T(y) + Z = C, \tag{8.29a}$$

$$A(X) = b, \tag{8.29b}$$

$$XZ = 0, \tag{8.29c}$$

$$X \succeq 0, \qquad Z \succeq 0. \tag{8.29d}$$

(The condition $XZ = 0$ in (8.29c) is a consequence of the definition (8.28) and the conditions $\mu = 0$ and (8.29d).)

The problem of minimizing the maximum eigenvalue of an affine matrix is an important application of SDP that arises in control theory, structural optimization, and other areas. Given matrices $T_0, T_1, \ldots, T_M$ in $\mathcal{S}^n$, the aim us to find scalars $z_1, \ldots, z_M$ such that the maximum eigenvalue of $T_0 + \sum_{k=1}^M z_k T_k$ is minimized. The SDP formulation is

$$\min_{(t,z) \in \mathbb{R}^{M+1}} t \qquad \text{subject to } T_0 + \sum_{k=1}^{M} z_k T_k - tI + Z = 0, \qquad Z \succeq 0. \tag{8.30}$$

Numerous other applications, including quadratically constrained QP, are discussed by Vandenberghe and Boyd [135]. Helmberg et al. [51] focus on applications to discrete optimization, showing that relaxations of graph-based problems such as max-cut, bisection, and maximum clique identification can be formulated as SDPs. Other applications are discussed by Alizadeh [3] and Boyd et al. [19].

Primal-dual algorithms for SDP are derived from the KKT conditions (8.29). As for linear programming, we define a nonlinear mapping F such that the primal-dual steps are Newton-like steps for F. The obvious extension to (1.5) is obtained by constructing F from the equality conditions in (8.29) and writing

$$F(X,y,Z) = \begin{bmatrix} A^T(y)+Z-C \\ A(X)-b \\ XZ \end{bmatrix} = 0. \tag{8.31}$$

The domain of the nonlinear mapping in this system (namely, $\mathcal{S}^n \times \mathbb{R}^m \times \mathcal{S}^n$) differs from its range $\mathcal{S}^n \times \mathbb{R}^m \times \mathbb{R}^{n\times n}$, since the product XZ is not symmetric in general. Hence, the ΔX component of a Newton-like step based on (8.31) is usually not symmetric, even when X and Z are symmetric. This observation has led researchers to reformulate the complementarity condition $XZ=0$ as an equivalent symmetric condition. Zhang [161] has proposed the condition $\mathcal{H}_P(XZ)=0$, where $\mathcal{H}_P$ is the symmetrization operator defined by

$$\mathcal{H}_P(M) = \tfrac{1}{2}(PMP^{-1}+(PMP^{-1})^T), \qquad P = \text{any nonsingular matrix}. \tag{8.32}$$

Newton-like steps $(\Delta X, \Delta y, \Delta Z)$ based on this condition, together with the primal and dual feasibility conditions from (8.26) and (8.27), are obtained by solving

$$\begin{bmatrix} A^T(\Delta y)+\Delta Z \\ A(\Delta X) \\ \mathcal{H}_P(\Delta XZ + X\Delta Z) \end{bmatrix} = \begin{bmatrix} -R_C \\ -r_b \\ -\mathcal{H}_P(XZ)+\sigma\mu I \end{bmatrix}, \tag{8.33}$$

where $\sigma \in [0,1]$ and

$$R_C = A^T(y)+\Delta Z - C, \qquad r_b = A(X)-b.$$

As in Framework PD (Chapter 1), we set

$$(X,y,Z) \leftarrow (X,y,Z)+\alpha(\Delta X, \Delta y, \Delta Z), \tag{8.34}$$

where the starting point (X^0, y^0, Z^0) and step length α are chosen so that $X > 0$ and $Z > 0$ at every iterate.

It is not completely straightforward to transform the step equation (8.33) into a system that can be solved by standard matrix factorizations, triangular substitutions, etc. The transformation depends strongly on the choice of matrix P in (8.32); Zhang [161] discusses the case of $P = Z^{1/2}$, whereas Alizadeh, Haeberly, and Overton [4] consider $P = I$.

Along with the Newton-like step, the other ingredients of primal-dual algorithms also extend to the SDP setting. The central path is the set of points $(X(\tau), y(\tau), Z(\tau))$, for $\tau > 0$, that solve

$$\min_{(X,y,Z)} \; -\log\det X - \log\det Z \quad \text{subject to}$$
$$A(X) = b, \quad A^T(y) + Z = C, \quad X \bullet Z = \tau, \quad X > 0, \quad Z > 0.$$

Neighborhoods of the central path are defined as

$$\mathcal{N}_2(\beta) = \{(X, y, Z) \,|\, X > 0,\, Z > 0,\, \|\mathcal{H}_P(XZ) - \mu I\|_F \leq \beta\mu\},$$
$$\mathcal{N}_{-\infty}(\gamma) = \{(X, y, Z) \,|\, X > 0,\, Z > 0,\, \text{eig}_i(XZ) \geq \gamma\mu, \text{all } i = 1, 2, \ldots, n\},$$

where $\|\cdot\|_F$ is the Frobenius norm and $\text{eig}_i(XZ)$ denotes the ith eigenvalue of XZ. (Note that the eigenvalues are all real because XZ is similar to the symmetric matrix $X^{1/2}ZX^{1/2}$.) The Tanabe–Todd–Ye potential function from Chapter 4 becomes

$$\Phi_\rho(X, Z) = \rho \log(\text{trace}\, XZ) - \log(\det XZ), \qquad \rho > n.$$

From these ingredients, we can construct analogues of the path-following, potential reduction, and infeasible-interior-point algorithms described in Chapters 4, 5, and 6. Mehrotra's heuristics (Chapter 10) extend likewise to the SDP setting, with similar success [4]. Superlinear convergence of a kind can also be attained, but the analysis is much more difficult than in the case of linear programming because of the complicated duality structure of the SDP.

Research into interior-point algorithms for SDP continues at a rapid pace. Some pointers to the current literature in this area can be found in the Notes and References section.

Proof of Theorem 8.4

This proof makes use of the KKT result, Theorem A.1.

Proof. Suppose that (x^*, Mx^*+q) solves the LCP. Then x^* is feasible for (8.14) since $x^* \geq 0$ and $Mx^*+q \geq 0$. By the complementarity condition, we have $(x^*)^T(Mx^*+q) = 0$. Because of the constraints in (8.14), all feasible points x must have $x^T(Mx+q) \geq 0$. Since x^* achieves this minimum possible value for the objective and is feasible, it solves (8.14), so we have proved one half of the result.

For the remaining part, suppose that x^* is a solution of (8.14). Since $x^T Mx = \frac{1}{2}x^T(M+M^T)x$, we can rewrite (8.14) equivalently as

$$\min_x \tfrac{1}{2}x^T(M+M^T)x + q^T x \quad \text{subject to } q+Mx \geq 0,\ x \geq 0. \tag{8.35}$$

Theorem A.1 implies that there are vectors $\bar{z}$ and $\hat{z}$ such that

$$\begin{aligned} q+(M+M^T)x^* - M^T\bar{z} - \hat{z} &= 0, \\ q+Mx^* &\geq 0, \\ (x^*, \bar{z}, \hat{z}) &\geq 0, \\ \bar{z}^T(Mx^*+q) &= 0, \\ \hat{z}^T x^* &= 0. \end{aligned}$$

Eliminating $\hat{z}$ from these conditions and rearranging slightly, we obtain

$$q+Mx^* \geq -M^T(x^*-\bar{z}), \tag{8.36a}$$
$$(x^*, \bar{z}) \geq 0, \tag{8.36b}$$
$$q+Mx^* \geq 0, \tag{8.36c}$$
$$\bar{z}^T(Mx^*+q) = 0, \tag{8.36d}$$
$$(x^*)^T(q+Mx^*+M^T(x^*-\bar{z})) = 0. \tag{8.36e}$$

Because $x_i^* \geq 0$ and $(q+Mx^*+M^T(x^*-\bar{z}))_i \geq 0$ for all $i = 1, 2, \ldots, n$, we have from (8.36e) that

$$x_i^*(q+Mx^*+M^T(x^*-\bar{z}))_i = 0, \qquad i = 1, 2, \ldots, n. \tag{8.37}$$

Since we also have $(q+Mx^*)_i \geq 0$ from (8.36c), it follows that

$$x_i^*(M^T(x^*-\bar{z}))_i \leq 0. \tag{8.38}$$

Since $\bar{z}_i \geq 0$, we can multiply the ith component of (8.36a) by $\bar{z}_i$ and use the complementarity condition (8.36d) to deduce that

$$-\bar{z}_i(M^T(x^*-\bar{z}))_i \leq \bar{z}_i(q+Mx^*)_i = 0. \tag{8.39}$$

Adding (8.38) and (8.39) and summing over i, we obtain

$$(x^* - \bar{z})^T M^T (x^* - \bar{z}) \le 0$$

and therefore

$$(x^* - \bar{z})^T M^T (x^* - \bar{z}) = 0$$

by positive semidefiniteness of M. Rearranging this equation and using (8.38) and (8.39) again, we obtain

$$0 \ge (x^*)^T M^T (x^* - \bar{z}) = \bar{z}^T M^T (x^* - \bar{z}) \ge 0$$

and therefore

$$(x^*)^T M^T (x^* - \bar{z}) = 0. \tag{8.40}$$

Summing over the components of (8.37) and using (8.40), we obtain

$$(x^*)^T (q + Mx^*) = 0.$$

Therefore, x^* is a solution of the LCP (8.1), and the proof is complete. □

Notes and References

The definitive text on LCPs is Cottle, Pang, and Stone [22]. The proof of Theorem 8.4 is from pages 138 and 139 of this book.

Another form of LCP, known as *geometric LCP*, is also equivalent to the standard LCP, a result that is proven in [10]. The class of problems known as P_* LCP is, on the other hand, a strict extension of monotone LCP. The form of P_* LCP problems is the same as (8.1), but the matrix M is allowed to have a slightly weaker property than positive semidefiniteness. Interior-point algorithms for such problems are identical to those for monotone LCP and their theoretical properties are similar; see, for example, Potra and Sheng [109].

The terms "degenerate" and "nondegenerate" are, unfortunately, used in different ways by different people. Following Mangasarian [82], we defined (above) a "degenerate LCP" to be one with no strictly complementary solutions and a "nondegenerate LCP" to be one for which such solutions do exist. Cottle, Pang, and Stone [22] do not use either of these terms but speak instead of "degenerate solutions," by which they mean solutions that are not strictly complementary.

Since feasible primal-dual linear programs always have strictly complementary solutions, they are always "nondegenerate" in the LCP sense. In linear programming, "primal degenerate" refers to problems (1.1) for which there are solutions x^* containing fewer than m nonzero components. "Dual degenerate" is defined analogously.

For asymptotic convergence results for interior-point methods on degenerate LCPs, see Monteiro and Wright [97]. This paper shows that methods whose steps eventually resemble affine-scaling steps *cannot* converge superlinearly on such problems. However, finite termination procedures analogous to those developed in Chapter 7 can be adapted to the LCP setting, and they obtain exact solutions for degenerate and nondegenerate problems alike.

Background information on SDP can be found in the book of Nesterov and Nemirovskii [101] and the excellent survey paper of Vandenberghe and Boyd [135]. The following papers contain descriptions and analyses of primal-dual algorithms: Alizadeh, Haeberly, and Overton [4]; Helmberg et al. [51]; Kojima, Shida, and Shindoh [70]; Kojima, Shindoh, and Hara [71]; Potra and Sheng [110]; and Zhang [161]. The best way to keep abreast of developments in SDP is to track new reports as they appear on Interior-Point Methods Online on the World Wide Web.

Exercises

1. Redefine the central path neighborhoods $\mathcal{N}_2(\theta)$ (1.15), $\mathcal{N}_{-\infty}(\gamma)$ (1.16), and $\mathcal{N}_{-\infty}(\gamma,\beta)$ (6.2) for the LCP setting.

2. Restate Algorithms SPF, PC, and LPF from Chapter 5 in terms of the LCP formulation.

3. Redefine the infeasible central path neighborhood $\mathcal{N}_{-\infty}(\gamma,\beta)$ in terms of the mLCP (8.4). Restate the infeasible-interior-point Algorithm IPF in terms of mLCP, using the step equations (8.8) as the basis of the algorithm. (*Note*: The algorithm designed here can be used to solve the simplified homogeneous self-dual problem of Chapter 9.)

4. Show that the coefficient matrix in (8.5) is positive semidefinite.

5. (i) Write the following convex QP in standard form (8.11):

$$\begin{aligned} &\min\ x^T H x + c^T x + d^T y \\ &\text{subject to } A_1 x + A_2 y \le b_1, \quad A_3 x = b_2, \quad y \le 0. \end{aligned} \tag{8.41}$$

(Hint: Split the variables x into positive and negative parts; that is, $x = x^+ - x^-$, $x^+ \ge 0$, $x^- \ge 0$.) Write this standard convex QP as an mLCP, as in (8.13).

(ii) Derive the KKT conditions for the problem (8.41), and state them as an mLCP. (Note: This mLCP contains fewer unknowns than the solution of the previous exercise.)

6. For the convex quadratic program (8.11) or its equivalent mLCP form (8.13), write the generalization of the Newton-like step equation (1.13).

7. Show that the properties described in Lemma 5.13 continue to hold for the LCP (8.1). That is, if (8.1) has a strictly complementary solution, then there is a constant C_3 such that for all $(x, s) \in \mathcal{N}_{-\infty}(\gamma)$, we have

$$\begin{aligned} 0 < x_i \le \mu/C_3 \quad (i \in \mathcal{N}), &\qquad 0 < s_i \le \mu/C_3 \quad (i \in \mathcal{B}), \\ s_i \ge C_3\gamma \quad (i \in \mathcal{N}), &\qquad x_i \ge C_3\gamma \quad (i \in \mathcal{B}), \end{aligned}$$

where $\mu = x^T s/n$.

8. Show that convexity of the functions g_i in (8.15) implies convexity of the feasible set. Show that the set of solutions to (8.15) is also convex.

9. Show that (8.30) has the dual form (8.27) of SDP by making the appropriate identifications between the variables and data.

Chapter 9

Detecting Infeasibility

All our analysis to this point has assumed that we are working with linear programs for which primal-dual solutions exist. Of course, not all linear programs have solutions. As we saw in Chapter 2, linear programs can be infeasible, or their objective functions can be unbounded below on the feasible region.

We know that primal-dual solutions exist whenever the primal and dual problems are both feasible (Theorem 2.1) and that the solution set is bounded when the set of strictly feasible points $\mathcal{F}^o$ is nonempty (Theorem 2.3). Hence, the very fact that we find a strictly feasible starting point for the algorithms of Chapters 4 and 5 guarantees that a solution exists.

In practice, though, we don't want to search for a strictly feasible starting point, since this is an expensive process in general. Most existing codes make use of a starting point (x^0, λ^0, s^0) that satisfies strict positivity $(x^0, s^0) > 0$ but not the equality conditions $Ax^0 = b$, $A^T\lambda^0 + s^0 = c$.

We have three options for handling an infeasible starting point. The first is to apply an infeasible-interior-point algorithm, such as Algorithm IPF from Chapter 6. We know from the convergence theory that when a solution exists, these algorithms will find it. When presented with an infeasible problem, these methods are unable to reduce the residuals below a certain (nonzero) level, so the iterates (x^k, s^k) diverge—their norm $\|(x^k, s^k)\|$ approaches ∞. This behavior is not particularly graceful, but practical codes can check for it and accurately diagnose infeasibility in most cases. There is some theoretical support for this approach: When $\|(x^k, s^k)\|$ becomes large, results due to Kojima, Megiddo, and Mizuno [63] can be used to identify a large subset of $\mathbb{R}^n_+ \times \mathbb{R}^m \times \mathbb{R}^n_+$ that does not contain any feasible points

$(x, \lambda, s) \in \mathcal{F}$. Details are given below.

The second way to deal with the starting point issue is to embed the given problem in a slightly larger problem for which a strictly feasible point is easy to find. The solution to the augmented problem should tell us everything we need to know about the original problem, including the following:

- If the original problem is infeasible, the augmented solution should tell us so. In fact, it should tell us whether the infeasibility occurs in the primal problem (2.1), the dual problem (2.2), or both.
- If the original problem *does* have a solution, we should be able to extract it easily from the augmented solution.

Since the augmented problem has a strictly feasible starting point, any of the "feasible" interior-point algorithms (see Chapters 4 and 5) can be used to solve it.

The *homogeneous self-dual (HSD)* formulation of Ye, Todd, and Mizuno [159] possesses most of these desirable properties and is quite efficient as well. Its cost per iteration is not much greater than the infeasible-interior-point approach applied to the original problem. A variant known as the *simplified HSD* formulation, due to Xu, Hung, and Ye [153], combines the infeasible-interior-point and HSD approaches. Like the HSD form, it always has a solution, even if the original problem does not. Unlike the HSD form, however, there are no strictly feasible starting points, so an infeasible-interior-point method must be used to solve the problem. Like the HSD, the simplified HSD yields either a solution of the original linear program (if one exists) or an indication of the infeasibility of the original problem.

We start this chapter by defining the term *self-dual*; further details of this intriguing property appear in Appendix A. We then describe the HSD and simplified HSD forms, and close with a discussion of the blowup test for detecting infeasibility in infeasible-interior-point methods.

Self-Duality

A self-dual linear program has the property that the primal and dual are actually *the same problem.* Obviously, this property can hold only if the formulation and the problem data are related in a special way. In the context of interior-point methods, self-duality is not just a theoretical curiosity—it allows us to eliminate half the equations from the KKT system and therefore to apply primal-dual algorithms in a particularly efficient way.

A self-dual linear program has the following general form:

$$\begin{array}{llll} & \min_{u,w} f^T u + g^T w & & \\ \text{subject to} & \begin{array}{rcl} M_{11}u + M_{12}w & \geq & -f, \\ -M_{12}^T u + M_{22}w & = & -g, \end{array} & u \geq 0,\ w \text{ free}, & (9.1) \end{array}$$

where the matrices M_{11} and M_{22} are square and skew-symmetric; that is, $M_{11} = -M_{11}^T$ and $M_{22} = -M_{22}^T$. Using a result from the exercises of Chapter 2 together with skew symmetry, we can verify that the dual of (9.1) is simply another copy of (9.1)! Not surprisingly, it follows that the KKT conditions for (9.1) contain a certain amount of redundancy. Eliminating this redundancy (see Appendix A), we can show that the solution to (9.1) can be derived from the solution of the following mixed monotone linear complementarity problem (mLCP):

$$\begin{bmatrix} v \\ 0 \end{bmatrix} = \begin{bmatrix} M_{11} & M_{12} \\ -M_{12}^T & M_{22} \end{bmatrix} \begin{bmatrix} u \\ w \end{bmatrix} + \begin{bmatrix} f \\ g \end{bmatrix}, \quad (u, v) \geq 0, \quad u^T v = 0. \quad (9.2)$$

Any solution (u, v, w) of (9.2) implies a solution (u, w) of (9.1). Conversely, any solution (u, w) of (9.1) implies a solution $(u, M_{11}u + M_{12}w + f, w)$ of (9.2). See Appendix A for details.

The mLCP (9.2) can be solved with an interior-point method, as discussed in Chapter 8. We recall from that chapter that mLCPs do not have strictly complementary solutions in general (that is, solutions (u, v, w) for which $u + v > 0$). (The possible absence of such solutions is a major point of difference between LCPs and linear programs; see Theorem 2.4.) Strict complementarity is needed for superlinear convergence of the algorithms of Chapter 7, so it is legitimate to ask whether such solutions exist for our *particular* mLCP(9.2). The answer is given in the following theorem, which is proven in Appendix A.

Theorem 9.1 *If the self-dual linear program* (9.1) *is feasible, the mLCP formulation* (9.2) *has a strictly complementary solution.*

The Simplified HSD Form

The simplified HSD formulation was first presented by Xu, Hung, and Ye [153]. It reformulates the original primal and dual pair (1.1), (1.2) as a single self-dual linear program and thence as an mLCP. The mLCP is guaranteed to have a solution, and it can be solved by an infeasible-interior-point method. The cost per iteration for solving the mLCP is only slightly greater than the cost per iteration for Algorithm IPF (Chapter 6) applied to the original primal-dual pair (1.1) and (1.2).

For convenience, we restate the original primal-dual pair:

$$\min c^T x \text{ subject to } Ax = b,\ x \geq 0, \tag{9.3a}$$

$$\max b^T \lambda \text{ subject to } A^T \lambda + s = c,\ s \geq 0. \tag{9.3b}$$

We now introduce a scalar variable τ and define the simplified HSD linear program as

$$\begin{array}{llcccl} \min & & & 0 & & \\ \text{subject to} & & -c^T x & +b^T \lambda & \geq & 0, \\ & c\tau & & -A^T \lambda & \geq & 0, \\ & -b\tau & +Ax & & = & 0, \\ & \multicolumn{5}{l}{\lambda \text{ free},\ (\tau, x) \geq 0.} \end{array} \tag{9.4}$$

It is easy to verify that (9.4) is self-dual: simply identify the data in (9.4) with (9.1) as follows:

$$M_{11} = \begin{bmatrix} 0 & -c^T \\ c & 0 \end{bmatrix}, \quad M_{12} = \begin{bmatrix} b^T \\ -A^T \end{bmatrix}, \quad M_{22} = 0,$$

$$f = 0, \quad g = 0, \quad u = (\tau, x), \quad w = \lambda.$$

The problem (9.4) is *homogeneous*, since the right-hand sides of all the constraints are zero. Homogeneity makes for a conical feasible region: If (τ, x, λ) is feasible for (9.4), the entire ray $\alpha(\tau, x, \lambda)$, $\alpha \geq 0$, is also feasible. Since the objective function in (9.4) is uniformly zero, all feasible points of this problem are also solutions.

As shown in the preceding section, the self-dual problem (9.4) gives rise to an equivalent mLCP formulation defined by

$$\begin{bmatrix} \kappa \\ s \\ 0 \end{bmatrix} = \begin{bmatrix} 0 & -c^T & b^T \\ c & 0 & -A^T \\ -b & A & 0 \end{bmatrix} \begin{bmatrix} \tau \\ x \\ \lambda \end{bmatrix}, \tag{9.5a}$$

$$(\tau, x, \kappa, s) \geq 0, \qquad \tau\kappa + x^T s = 0, \tag{9.5b}$$

where $\kappa \in \mathbb{R}$ and $s \in \mathbb{R}^n$ are the surplus variables for the inequality constraints in (9.4). We can now solve this problem with an infeasible-interior-point method such as Algorithm IPF of Chapter 6, modified for mLCP as described in Chapter 8.

Theorems 9.2 and 9.3 show that we need a strictly complementary solution for (9.5) to recover information about the feasibility of the original primal and dual problems (9.3). Such a solution $(\tau^*, x^*, \lambda^*, \kappa^*, s^*)$ must satisfy

$$\tau^* + \kappa^* > 0, \qquad x^* + s^* > 0.$$

Existence of strictly complementary solutions for (9.5) follows immediately from Theorem 9.1, since the corresponding linear program (9.4) is obviously feasible.

Fortunately, path-following algorithms for linear programming approach only strictly complementary limit points, as we showed in Lemma 5.13 and Theorems 5.14 and 6.8. This property continues to hold when we extend these algorithms to mLCPs—provided, of course, that strictly complementary solutions exist for the mLCP.

On the other hand, strict feasibility is not satisfied by the problem (9.4). If a strictly feasible point $(\bar{\tau}, \bar{x}, \bar{\lambda})$ existed, it would satisfy

$$b^T\bar{\lambda} > c^T\bar{x}, \qquad A^T\bar{\lambda} < \bar{\tau}c, \qquad A\bar{x} = b, \qquad (\bar{\tau}, \bar{x}) > 0. \tag{9.6}$$

Hence, the point (x, λ, s) defined by

$$(x, \lambda, s) = (\bar{x}/\bar{\tau}, \bar{\lambda}/\bar{\tau}, c - A^T\bar{\lambda}/\bar{\tau})$$

would be a strictly feasible primal-dual point for (9.3). By the duality results of Chapter 2 (in particular, the inequality (2.7)), we obtain

$$0 < x^Ts = c^Tx - b^T\lambda = \frac{1}{\bar{\tau}}(c^T\bar{x} - b^T\bar{\lambda}) < 0,$$

where the last inequality follows from (9.6). This inequality makes no sense, so we conclude that no such point $(\bar{\tau}, \bar{x}, \bar{\lambda})$ can exist.

We complete this section by showing how the primal-dual solution set Ω of (9.3) is related to the solution set for (9.2). In the infeasible case $\Omega = \emptyset$, we can use the solution of (9.5) to deduce which of the two problems (9.3a) and (9.3b) is infeasible. In the feasible case (Ω nonempty), we show how a solution of (9.3) can be recovered from a solution of (9.5).

The first result deals with the feasible case.

Theorem 9.2 *The primal-dual solution set Ω for* (9.3) *is nonempty if and only if all strictly complementary solutions $(\tau^*, x^*, \lambda^*, \kappa^*, s^*)$ of* (9.5) *have $\tau^* > 0$ and $\kappa^* = 0$.*

Proof. Given a strictly complementary solution of (9.5) with $\tau^* > 0$ and $\kappa^* = 0$, the point (x, λ, s) defined by

$$(x, \lambda, s) = \frac{1}{\tau^*}(x^*, \lambda^*, s^*)$$

satisfies $Ax = b$, $A^T\lambda + s = c$, $(x, s) \geq 0$, and

$$x^T s = \frac{1}{\tau^*}(x^*)^T(c - A^T\lambda^*/\tau^*) = \frac{1}{\tau^*}(c^T x^* - b^T\lambda^*) = \frac{\kappa^*}{\tau^*} = 0.$$

Therefore (x, λ, s) is a primal-dual solution for (9.3).

Conversely, suppose that the primal-dual solution set Ω for (9.3) is nonempty. Because of Theorem 2.4, Ω contains strictly complementary solutions. Let (x, λ, s) be such a solution, and define

$$(\tau^*, x^*, \lambda^*, \kappa^*, s^*) = (1, x, \lambda, 0, s).$$

It is easy to check that this point is a solution of (9.2). In fact, it is a strictly complementary solution, since

$$(\tau^*, x^*) + (\kappa^*, s^*) = (1 + 0, x + s) > 0.$$

Since the locations of the zeros and nonzero elements in the vectors (τ^*, x^*) and (κ^*, s^*) are the same in all strictly complementary solutions, we have $\tau^* > 0$ and $\kappa^* = 0$ for all such solutions, as claimed. □

If we find a strictly complementary solution of (9.5) for which $\tau^* = 0$, then one or both of the problems (9.3) is infeasible. The last result in this section shows how we can glean information from a solution of this kind.

Theorem 9.3 *Suppose that* (9.5) *has a strictly complementary solution* $(\tau^*, x^*, \lambda^*, \kappa^*, s^*)$ *for which* $\kappa^* > 0$. *Then at least one of* $c^T x^*$ *and* $-b^T\lambda^*$ *is negative and*

- *if* $c^T x^* < 0$, *the dual problem* (9.3b) *is infeasible;*
- *if* $-b^T\lambda^* < 0$, *the primal problem* (9.3a) *is infeasible.*

Proof. Since $\kappa^* > 0$, its complementary variable τ^* must be zero. It follows from (9.5a) that at least one of $c^T x^*$ and $-b^T\lambda^*$ is negative.

Suppose that $c^T x^* < 0$. Since $\tau^* = 0$, we have from (9.5) that $x^* \geq 0$ and $Ax^* = b\tau^* = 0$. If the dual problem (9.3b) were feasible, we would have a vector $(\bar{\lambda}, \bar{s})$ with $\bar{s} \geq 0$ and $A^T\bar{\lambda} + \bar{s} = c$. By multiplying the latter equation by $(x^*)^T$, we obtain

$$0 \geq c^T x^* = \bar{\lambda}^T A x^* + \bar{s}^T x^* = \bar{s}^T x^* \geq 0,$$

which is a contradiction. The proof of the second claim is similar. □

Implementations of the infeasible-interior-point method usually take different step lengths in the primal and dual variables, as we discuss in Chapter 10. It is not obvious how this feature can be implemented in the simplified HSD form of this section, since the reformulation causes the primal and dual variables to be coupled. Xu, Hung, and Ye [153] show that different step lengths α^{pri} and α^{dual} can be used for x and s components, respectively, while a common step length of *either* α^{pri} *or* α^{dual} can be used for τ and κ, according to which of these two alternatives gives the smaller value of τ.

The HSD Form

The HSD form described by Ye, Todd, and Mizuno [159] is slightly more complicated than the simplified variant (9.4), since it incorporates an extra row and column in the constraints to account for the infeasibility of the chosen initial point with respect to the original problem pair (9.3). The added complication yields an advantage, however: It becomes easy to choose a strictly feasible initial point, so we are no longer restricted to using an infeasible-interior-point algorithm to solve the problem. Strict feasibility also implies that the solution set is bounded, a property that improves the numerical behavior of the implementation.

Given initial values for the primal-dual vector (x^0, λ^0, s^0) satisfying $x^0 > 0$ and $s^0 > 0$, the HSD form is defined as

$$\begin{array}{lrrrrcl}
\min & & & ((x^0)^T s^0 + 1)\theta & & & \\
\text{subject to} & & -c^T x & +b^T\lambda & +\bar{z}\theta & \geq & 0, \\
& c\tau & & -A^T\lambda & -\bar{c}\theta & \geq & 0, \\
& -b\tau & +Ax & & +\bar{b}\theta & = & 0, \\
& -\bar{z}\tau & +\bar{c}^T x & -\bar{b}^T\lambda & & = & -((x^0)^T s^0 + 1), \\
& & \multicolumn{3}{l}{(\lambda, \theta) \text{ free}, \ (\tau, x) \geq 0,} & &
\end{array} \tag{9.7}$$

where

$$\bar{b} = b - Ax^0, \qquad \bar{c} = c - A^T\lambda^0 - s^0, \qquad \bar{z} = c^T x^0 + 1 - b^T\lambda^0.$$

The vectors $\bar{b}$ and $\bar{c}$ represent infeasibility of the chosen initial vector (x^0, λ^0, s^0) with respect to the constraints of the primal-dual pair. The scalar $\tau \in \mathbb{R}$ plays the same role as in (9.4), whereas $\theta \in \mathbb{R}$ scales the initial infeasibility vectors. We show in Theorem 9.4 below that $\theta = 0$ at all solutions of (9.7).

Self-duality of (9.7) is easy to check. Note that the last constraint in (9.7) has a nonzero right-hand side, so the problem is not, strictly speaking, homogeneous. This constraint serves only to normalize the solution components (s, λ, τ), however, so we don't let its presence affect the terminology.

For the equivalent mLCP formulation of (9.7), we introduce surplus variables $\kappa \in \mathbb{R}$ and $s \in \mathbb{R}^n$ for the first and second constraints, respectively. As in (9.2), we obtain

$$\begin{bmatrix} \kappa \\ s \\ 0 \\ 0 \end{bmatrix} = \begin{bmatrix} 0 & -c^T & b^T & \bar{z} \\ c & 0 & -A^T & -\bar{c} \\ -b & A & 0 & \bar{b} \\ -\bar{z} & \bar{c}^T & -\bar{b}^T & 0 \end{bmatrix} \begin{bmatrix} \tau \\ x \\ \lambda \\ \theta \end{bmatrix} + \begin{bmatrix} 0 \\ 0 \\ 0 \\ (x^0)^T s^0 + 1 \end{bmatrix}, \tag{9.8a}$$

$$(\tau, x, \kappa, s) \geq 0, \qquad \tau\kappa + x^T s = 0. \tag{9.8b}$$

A strictly feasible initial point for this problem—one that satisfies (9.8a) together with $(\tau, x, \kappa, s) > 0$—is given by

$$(\tau, x, \lambda, \theta, \kappa, s) = (1, x^0, \lambda^0, 1, 1, s^0).$$

Note that $(\tau, x, \lambda, \theta) = (1, x^0, \lambda^0, 1)$ is a strictly feasible point for the linear programming form (9.7). It follows from Theorem 2.3 together with self-duality that the solution set for (9.7) is bounded. This fact, in turn, implies boundedness for the solution set of the mLCP form (9.8).

As mentioned above, the θ component of any solution of the HSD form is zero.

Theorem 9.4 *If* $(\tau^*, x^*, \lambda^*, \theta^*, \kappa^*, s^*)$ *is any solution of* (9.8), *then* $\theta^* = 0$.

Proof. If $(\tau, x, \lambda, \theta, \kappa, s)$ is any feasible point for (9.8), we have

$$\begin{aligned} 0 &\leq x^T s + \tau\kappa \\ &= x^T(-A^T\lambda + c\tau - \bar{c}\theta) + \tau(b^T\lambda - c^T x + \bar{z}\theta) \\ &= \lambda^T(-Ax + b\tau) + \theta(-\bar{c}^T x + \bar{z}\tau) \\ &= \theta(\bar{b}^T\lambda - \bar{c}^T x + \bar{z}\tau) \\ &= \theta((x^0)^T s^0 + 1). \end{aligned}$$

At a solution $(\tau^*, x^*, \lambda^*, \theta^*, \kappa^*, s^*)$, we have $0 = (x^*)^T s^* + \tau^*\kappa^*$. Hence, since $(x^0)^T s^0 + 1 > 0$ it follows that $\theta^* = 0$, as claimed. □

The property $\theta^* = 0$ means that the last column in the constraint set in (9.7) vanishes at all solutions, so the constraints look more and more like the constraints for the simplified HSD form (9.4) as we approach the solution set. Not surprisingly, then, the main theoretical results for simplified HSD—Theorems 9.2 and 9.3—also hold for the HSD form. We state these results and leave the proofs as exercises, since they differ only slightly from the earlier proofs.

Theorem 9.5 *The primal-dual solution set* Ω *for* (9.3) *is nonempty if and only if all strictly complementary solutions* $(\tau^*, x^*, \lambda^*, \theta^*, \kappa^*, s^*)$ *of* (9.8) *have* $\tau^* > 0$ *and* $\kappa^* = 0$.

Theorem 9.6 *Suppose that* (9.8) *has a strictly complementary solution* $(\tau^*, x^*, \lambda^*, \theta^*, \kappa^*, z^*)$ *for which* $\kappa^* > 0$. *Then at least one of* $c^T x^*$ *and* $-b^T \lambda^*$ *is negative, and*

- *if* $c^T x^* < 0$, *the dual problem* (9.3b) *is infeasible;*
- *if* $-b^T \lambda^* < 0$, *the primal problem* (9.3a) *is infeasible.*

Solutions of the primal-dual pair (9.3) can be recovered from solutions for (9.8) in exactly the same way as for simplified HSD.

Identifying a Solution-Free Region

In the description and analysis of Algorithm IPF in Chapter 6, we assumed that the primal-dual feasible set $\mathcal{F}$, and hence the solution set Ω, was nonempty. An obvious stopping criterion for this algorithm, which we have used already in proving the polynomial complexity result, Theorem 6.2, is

$$\mu_k \leq \epsilon, \tag{9.9}$$

where ϵ is some small positive constant. Since the norm of the residual vector (r_c^k, r_b^k) is bounded by a multiple of μ_k (see (6.2)), it follows from (9.9) that the infeasibility is also small and hence that the current point is close to optimality. We showed in Theorem 6.1 that whenever $\Omega \neq \emptyset$, we have $\mu_k \downarrow 0$, so the criterion (9.9) is guaranteed to hold after a finite number of iterations.

If, on the other hand, there are no solutions, the norm of (x^k, s^k) increases to ∞ as $k \to \infty$. The proof of this result and the other results in this section is adapted from Kojima, Megiddo, and Mizuno [63].

Theorem 9.7 *If* $\Omega = \emptyset$, *the sequence* $\{(x^k, \lambda^k, s^k)\}$ *generated by Algorithm IPF has*

$$\lim_{k \to \infty} \|(x^k, s^k)\| = \infty.$$

Proof. We prove the result only for the case in which A has full row rank and leave the general case as an exercise. We reuse some of the analysis of Chapter 6 during this proof.

Suppose for contradiction that there is an infinite subsequence $\mathcal{K}$ and a constant ω such that

$$\|(x^k, s^k)\|_1 \leq \omega \qquad \text{for all } k \in \mathcal{K}.$$

Because of compactness, there must be a subsequence $\bar{\mathcal{K}} \subset \mathcal{K}$ and a point $(\bar{x}, \bar{s})$ such that

$$\lim_{k \in \bar{\mathcal{K}}} (x^k, s^k) = (\bar{x}, \bar{s}).$$

Recall that μ_k is monotonically decreasing. If $\mu_k \downarrow 0$, we have from $(x^k, \lambda^k, s^k) \in \mathcal{N}_{-\infty}(\gamma, \beta)$ that $\|r_b^k\| \to 0$ and $\|r_c^k\| \to 0$. Hence for the limit point $(\bar{x}, \bar{s})$, we have $\bar{x}^T \bar{s} = 0$, $A\bar{x} = b$, and $A^T \bar{\lambda} + \bar{s} = c$ for some $\bar{\lambda}$. But then $(\bar{x}, \bar{\lambda}, \bar{s})$ is a primal-dual solution, which contradicts $\Omega = \emptyset$.

Hence, $\{\mu_k\}$ cannot approach zero, so there is a value $\bar{\mu} > 0$ such that $\mu_k \geq \bar{\mu}$ for all k. We define a set $\mathcal{G}^*$ that contains all such points by

$$\mathcal{G}^* = \left\{ (x, \lambda, s) \in \mathcal{N}_{-\infty}(\gamma, \beta) \,|\, x^T s/n \geq \bar{\mu}, \ \ \|(x, s)\|_1 \leq \omega \right\}.$$

Note that $\mathcal{G}^*$ is compact and that $(x^k, \lambda^k, s^k) \in \mathcal{G}^*$ for all $k \in \mathcal{K}$. The solution $(\Delta x, \Delta \lambda, \Delta s)$ of the step equations (1.20) is unique in its Δx and Δs components for all $(x, \lambda, s) \in \mathcal{G}^*$ and all centering parameters $\sigma \in [0, 1]$. Therefore, by compactness, we can define a uniform upper bound $\hat{\omega} > 0$ on the step vector $(\Delta x, \Delta s)$ for all $(x, \lambda, s) \in \mathcal{G}^*$ and $\sigma \in [0, 1]$; that is,

$$\|(\Delta x^k, \Delta s^k)\| \leq \hat{\omega} \qquad \text{for all } k \in \mathcal{K} \text{ and some } \hat{\omega} > 0.$$

We can now use this step bound as the basis of a similar argument to the proof of Lemma 6.7, to show that there is a lower bound on the step length—a constant $\bar{\alpha} > 0$ such that $\alpha_k \geq \bar{\alpha}$ for all $k \in \mathcal{K}$. From the sufficient decrease condition (6.6), it follows that

$$\mu_{k+1} \leq (1 - .01\alpha_k)\mu_k \leq (1 - .01\bar{\alpha})\mu_k \qquad \text{for all } k \in \mathcal{K}.$$

Because $\{\mu_k\}$ is monotonic and $\mathcal{K}$ is an infinite sequence, it follows immediately that $\mu_k \downarrow 0$. This limit contradicts $\mu_k \geq \bar{\mu}$, so our proof is complete. □

Theorem 9.7 motivates the following simple termination test for Algorithm IPF: For some positive constants ϵ (small) and ω (large), terminate the algorithm if

$$\mu_k \leq \epsilon \qquad \text{or} \qquad \|(x^k, s^k)\|_1 \geq \omega. \tag{9.10}$$

From the discussion above, we know that one of the two conditions in (9.10) eventually holds for some k sufficiently large, so the algorithm is guaranteed to terminate after a finite number of iterations.

If termination happens because of the first condition in (9.10), we are close to an approximate solution. If termination occurs because of the second condition in (9.10), chances are that Ω and hence $\mathcal{F}$ is empty. Since ω is a finite number, we can't be *certain* that $\mathcal{F}$ is empty, but we *can* define a large set (which we call $\bar{\mathcal{F}}$) that does not intersect with $\mathcal{F}$. The set $\bar{\mathcal{F}}$ is defined in terms of two positive parameters $\bar{\delta}$ and $\bar{\omega}$ that are assumed to satisfy the condition

$$\frac{\bar{\omega}^2 + n\mu_0}{\bar{\delta}} \leq \omega. \tag{9.11}$$

(Note that $(\bar{\delta}, \bar{\omega}) > 0$ is usually not uniquely defined by (9.11).) Now, defining the set $\bar{\mathcal{F}}$ by

$$\bar{\mathcal{F}} = \left\{ (\bar{x}, \bar{\lambda}, \bar{s}) \,|\, \bar{x} \geq \bar{\delta}e, \;\; \bar{s} \geq \bar{\delta}e, \;\; \|(\bar{x}, \bar{s})\|_1 \leq \bar{\omega} \right\}, \tag{9.12}$$

we have the following result.

Theorem 9.8 *Suppose that the starting point for Algorithm IPF lies inside $\bar{\mathcal{F}}$ and that the algorithm terminates at some iterate $k > 0$ with $\|(x^k, s^k)\|_1 \geq \omega$. Then $\bar{\mathcal{F}}$ contains no primal-dual feasible vectors; that is, $\bar{\mathcal{F}} \cap \mathcal{F} = \emptyset$.*

Proof. We prove the result by contradiction, showing that if there exists a feasible point inside $\bar{\mathcal{F}}$, the condition (9.11) must be violated.

Recall that for ν_k defined by (6.7) and r_b and r_c defined by (6.1), we have from (6.8) that

$$Ax^k - b = r_b^k = \nu_k r_b^0, \qquad A^T\lambda^k + s^k - c = r_c^k = \nu_k r_c^0. \tag{9.13}$$

Suppose that we have a point (x, λ, s) that lies in $\mathcal{F} \cap \bar{\mathcal{F}}$, and define another point $(\bar{x}, \bar{\lambda}, \bar{s})$ by

$$(\bar{x}, \bar{\lambda}, \bar{s}) = (1 - \nu_k)(x, \lambda, s) + \nu_k(x^0, \lambda^0, s^0).$$

Since (x, λ, s) and (x^0, λ^0, s^0) both belong to the convex set $\bar{\mathcal{F}}$, so does $(\bar{x}, \bar{\lambda}, \bar{s})$, since it lies on a line between these two points.

Because of (9.13) and feasibility of (x, λ, s), we have that

$$A\bar{x} - b = \nu_k r_b^0, \qquad A^T\bar{\lambda} + \bar{s} - c = \nu_k r_c^0$$

so that $(\bar{x}, \bar{\lambda}, \bar{s})$ satisfies the same equations as (x^k, λ^k, s^k) in (9.13). Hence, by a standard argument (see, for instance, the proof of Lemma 2.5), we have

$$(\bar{x} - x^k)^T(\bar{s} - s^k) = 0. \tag{9.14}$$

Therefore we have

$$\begin{array}{rcll} \bar{\omega}^2 + n\mu_0 & \geq & \bar{x}^T\bar{s} + (x^k)^T s^k & \text{since } \|(\bar{x}, \bar{s})\|_1 \leq \bar{\omega} \text{ and } (x^k)^T s^k < n\mu_0 \\ & = & \bar{x}^T s^k + \bar{s}^T x^k & \text{by (9.14)} \\ & \geq & \bar{\delta} e^T s^k + \bar{\delta} e^T x^k & \text{since } \bar{x} \geq \bar{\delta} e \text{ and } \bar{s} \geq \bar{\delta} e \\ & = & \bar{\delta} \|(x^k, s^k)\|_1 & \\ & \geq & \bar{\delta}\omega & \text{since } \|(x^k, s^k)\|_1 \geq \omega \text{ at termination.} \end{array}$$

Since this inequality contradicts (9.11), we conclude that $\mathcal{F} \cap \bar{\mathcal{F}}$ must be empty. □

We can make the region $\bar{\mathcal{F}}$ as extensive as we like by choosing ω sufficiently large, since from (9.11), $\bar{\delta}$ can be made arbitrarily small and $\bar{\omega}$ arbitrarily large. The set $\bar{\mathcal{F}}$ is a little unsatisfying; since $\bar{\delta} > 0$, it always excludes a neighborhood of the origin. Kojima, Megiddo, and Mizuno [63] suggest a modified algorithm in which the residuals r_b^k and r_c^k are "frozen" once they fall below a certain threshold. They are replaced by zeros on the right-hand sides of the step equations (1.20). If this algorithm terminates due to $\|(x^k, s^k)\| \geq \omega$, there is a no feasible point inside a set of the form

$$\{(\bar{x}, \bar{\lambda}, \bar{s}) \mid (\bar{x}, \bar{s}) \geq 0, \quad \|(\bar{x}, \bar{s})\|_1 \leq \bar{\omega}\}$$

for some $\bar{\omega} > 0$. However, Lustig, Marsten, and Shanno [78] report on computational experience with this strategy and conclude that is it not reliable on difficult problems.

Implementations of the HSD Formulations

We see from the analysis of the three approaches—infeasible-interior-point, HSD, and simplified HSD—that the two HSD formulations handle infeasible linear programs in a cleaner way than the infeasible method. There is a computational price to pay for this advantage, however. It costs more to solve the linear systems at each iteration of an interior-point method for the mLCPs (9.5) and (9.8) than it does to solve the system (1.20) at each iteration of the infeasible-interior-point method. Still, as we now show, the extra cost is not very significant if we make use of the Sherman–Morrison–Woodbury formula (see (A.21) and (A.24) in Appendix A).

Suppose that we apply an interior-point method to the mLCP for the simplified HSD form (9.5). From (8.8), we see that the linear system to be solved at each iteration has the following form:

$$\begin{bmatrix} 0 & -c^T & b^T & -1 & 0 \\ c & 0 & -A^T & 0 & -I \\ -b & A & 0 & 0 & 0 \\ \kappa & 0 & 0 & \tau & 0 \\ 0 & S & 0 & 0 & X \end{bmatrix} \begin{bmatrix} \Delta\tau \\ \Delta x \\ \Delta\lambda \\ \Delta\kappa \\ \Delta s \end{bmatrix} = \begin{bmatrix} -r_\tau \\ -r_c \\ -r_b \\ -\kappa\tau + \sigma\mu \\ -XSe + \sigma\mu e \end{bmatrix}, \tag{9.15}$$

where $\mu = (x^T s + \kappa\tau)/(n+1)$, $\sigma \in [0,1]$, and r_τ, r_c, and r_b are residual vectors. Eliminating the two scalar variables from this system of equations and applying a sign change to one of the rows, we obtain

$$\left(\begin{bmatrix} 0 & A^T & I \\ A & 0 & 0 \\ S & 0 & X \end{bmatrix} + \frac{\tau}{\kappa} \begin{bmatrix} -c \\ -b \\ 0 \end{bmatrix} \begin{bmatrix} c \\ -b \\ 0 \end{bmatrix}^T \right) \begin{bmatrix} \Delta x \\ \Delta\lambda \\ \Delta s \end{bmatrix} = \begin{bmatrix} -\bar{r}_c \\ -\bar{r}_b \\ -XSe + \sigma\mu e \end{bmatrix}, \tag{9.16}$$

where $\bar{r}_c$ and $\bar{r}_b$ are defined appropriately. In contrast, the linear system that is solved at each iteration of the infeasible-interior-point iteration for (9.3) has the general form

$$\begin{bmatrix} 0 & A^T & I \\ A & 0 & 0 \\ S & 0 & X \end{bmatrix} \begin{bmatrix} \Delta x \\ \Delta\lambda \\ \Delta s \end{bmatrix} = \begin{bmatrix} -r_c \\ -r_b \\ -XSe + \sigma\mu e \end{bmatrix}. \tag{9.17}$$

The coefficient matrices in (9.16) and (9.17) differ by a rank-one term. However, the matrix in (9.16) will usually be denser, since the vectors c and b usually have a much higher proportion of nonzeros than the matrix A. It would be naive to factor the matrix in (9.16) directly and ignore its relationship to the sparser matrix (9.17). A much smarter idea is to use the formula (A.22), which says that the solution y of the linear system

$$(\tilde{G} + ab^T)y = d$$

satisfies

$$y = \tilde{G}^{-1}d - \frac{b^T(\tilde{G}^{-1}d)}{1 + b^T(\tilde{G}^{-1}a)}\tilde{G}^{-1}a.$$

This formula implies that we can solve (9.16) by first solving the two linear systems

$$\begin{bmatrix} 0 & A^T & I \\ A & 0 & 0 \\ S & 0 & X \end{bmatrix} \begin{bmatrix} y_1 \\ y_2 \\ y_3 \end{bmatrix} = \begin{bmatrix} -c \\ -b \\ 0 \end{bmatrix}, \tag{9.18a}$$

$$\begin{bmatrix} 0 & A^T & I \\ A & 0 & 0 \\ S & 0 & X \end{bmatrix} \begin{bmatrix} y_4 \\ y_5 \\ y_6 \end{bmatrix} = \begin{bmatrix} \bar{r}_c \\ \bar{r}_b \\ -XSe + \sigma\mu e \end{bmatrix} \tag{9.18b}$$

and then performing some elementary vector operations. The two systems (9.18) can be solved by performing a single factorization on the common coefficient matrix, followed by two back-substitutions, one for each of the right-hand sides. Since the computational cost of a factorization is significantly higher than the cost of back-substitution, the total cost of solving the pair of systems is not much greater than the cost of solving a single system in isolation.

For the original HSD form (9.8), each iteration of the interior-point method requires the solution of *three* linear systems like the ones in (9.18). Again, the coefficient matrix is the same for each system, so we need a single factorization operation and three substitution operations. The details are left as an exercise.

Notes and References

In the discussion above, we applied the infeasibility detection results of Kojima, Megiddo, and Mizuno [63] to the specific Algorithm IPF from Chapter 6. It is trivial to state these results so that they apply to a wide range of algorithms, as is done in [63]. The analysis requires only that the iterates satisfy some essential property such as membership of the neighborhood $\mathcal{N}_{-\infty}(\gamma, \beta)$.

Xu, Hung, and Ye [153] describe an experimental code that is based on their simplified HSD formulation, and give numerical results. Performance of the code on both feasible and infeasible problems from the Netlib collection is strong.

Exercises

1. Using the skew-symmetric property of A, verify that the mLCP (9.2) is monotone.

2. Verify that the problem (9.7) is self-dual.

3. Modify the proof of Theorem 9.7 for the case in which A may not have full row rank.

4. Prove Theorems 9.5 and 9.6 by modifying the proofs of Theorems 9.2 and 9.3.

5. For the linear system (9.15) corresponding to the mLCP (9.5), derive the reduced form (9.16), including exact definitions of $\bar{r}_c$ and $\bar{r}_b$.

6. For the mLCP arising from the HSD formulation (9.8), show that the linear system at each interior-point iteration can be reduced to an equation similar to (9.16), except that the perturbation term has rank two rather than rank one.

Chapter 10

Practical Aspects of Primal-Dual Algorithms

Most interior-point codes have since 1990 been based on Mehrotra's predictor-corrector algorithm. Mehrotra's algorithm departs slightly from the primal-dual Framework PD in that it enhances the basic Newton-like search direction with a correction that is fairly inexpensive to compute. It also allows an adaptive choice of the centering parameter σ at each iteration. The algorithm builds on the theory of Chapters 4, 5, and 6, together with some other ideas from optimization and numerical analysis that we describe below. It also incorporates a number of heuristics that have been developed during ten years of computational experience.

We begin this chapter by motivating Mehrotra's approach and defining Algorithm MPC, an implementation of this approach. (There are several variants of Mehrotra's algorithm, all quite similar.) We then examine two methods—Shamanskii's method for nonlinear equations, and a higher-order method for trajectory following—that illustrate nicely the fundamental ideas on which Mehrotra's algorithm is based. We close with a discussion of several enhancements of the algorithm that have been implemented in published software.

Mehrotra's approach was not cut from whole cloth, as its author readily admits in [88]. One of its primary features—the use of higher-order approximations to the central path—was foreshadowed by Megiddo [85] and further developed by Monteiro, Adler, and Resende [95] and Karmarkar et al. [58]. Another key ingredient—the infeasible-interior-point path-following approach—had already been implemented with great success by Lustig,

Marsten, and Shanno [77]. Mehrotra's contribution was to combine these existing ideas in just the right way and to add ingenious heuristics for choosing the centering parameter (adaptively), the step lengths, and a starting point. The result was a highly efficient algorithm that, as we have noted, still forms the basis of most existing interior-point codes.

Motivation for Mehrotra's Algorithm

Mehrotra's algorithm generates a sequence of infeasible iterates (x^k, λ^k, s^k) for which $(x^k, s^k) > 0$. The search direction at each iteration consists of three components:

- an affine-scaling "predictor" direction—the pure Newton direction for the function $F(x, \lambda, s)$ defined by (1.5a),
- a centering term whose size is governed by the adaptively chosen centering parameter σ,
- a "corrector" direction that attempts to compensate for some of the nonlinearity in the affine-scaling direction.

The first two components—the affine-scaling step and the centering term—combine to make up the standard infeasible-interior-point step from the generic step equations (1.20). In Mehrotra's algorithm, however, the affine-scaling component is calculated separately from, and prior to, the centering component. By arranging the computations in this way, we gain a key advantage: the ability to choose the centering parameter σ *adaptively* rather than a priori, as in the algorithms of Chapters 4, 5, and 6. If the affine-scaling direction makes good progress in reducing the duality measure μ while remaining inside the positive orthant defined by $(x, s) > 0$, we conclude that little centering is needed at this iteration, so we assign a small value to σ. If, on the other hand, we can move only a short distance along the affine-scaling direction before violating the constraints $(x, s) > 0$, we conclude that a significant amount of centering is needed, so we choose σ closer to 1.

The drawback of calculating the centering component separately in this fashion is that we need to solve *two* linear systems instead of one at each iteration. However, the marginal cost is not very great. Since the two systems have the same coefficient matrix, we need only factor the matrix once and perform two back-substitutions, one for each right-hand side.

The affine-scaling direction is obtained from a linear approximation to the equality conditions in the KKT system (1.4)—the pure Newton direction for the function $F(x, \lambda, s)$ from (1.5a). Since this direction is computed separately, it can be used to assess the error in the linear approximation. Knowledge of this error allows us to calculate a *corrector* component—the third component of the Mehrotra search direction—and therefore improve our linear, first-order model of F to a quadratic, second-order model. Since the centering and corrector components are obtained by solving linear systems with the same coefficient matrix as the affine-scaling direction, and since they are independent of each other, there is no need to compute them separately. We can simply merge them into a single direction by adding their corresponding right-hand sides and compute the combined direction with a single back-substitution.

For the cost of a single extra back-substitution, Mehrotra's algorithm offers an adaptive choice of σ and a higher-order enhancement of the basic pure Newton/affine-scaling step. In general, the extra cost per iteration is easily justified by a significant reduction in the number of iterations.

In our description of Mehrotra's algorithm, we take different step lengths in the primal and dual variables, departing from the convention of a single common step length that we used in Chapters 4, 5, and 6. The idea of different primal and dual step lengths did not originate with Mehrotra—it has been used in both theoretical and practical development of primal-dual methods since the earliest days. Computationally, this enhancement saves about 10%–20% on a typical problem. We used a single common step length in the earlier chapters mainly for reasons of convenience, to avoid complication in the analysis. A common step length also has the benefit of generality: Algorithms for QP, LCP, and the like are constrained to use a common step because of tighter coupling of their variables (see Chapter 8).

The Algorithm

We start by outlining a single iteration and then define the method. Throughout the chapter, we drop the iteration counter k from matrix and vector quantities.

Given a point (x, λ, s) with $(x, s) > 0$, the affine-scaling direction is found by solving the following system:

$$\begin{bmatrix} 0 & A^T & I \\ A & 0 & 0 \\ S & 0 & X \end{bmatrix} \begin{bmatrix} \Delta x^{\text{aff}} \\ \Delta \lambda^{\text{aff}} \\ \Delta s^{\text{aff}} \end{bmatrix} = \begin{bmatrix} -r_c \\ -r_b \\ -XSe \end{bmatrix}, \tag{10.1}$$

where the residuals are defined as usual by

$$r_b = Ax - b, \qquad r_c = A^T\lambda + s - c. \tag{10.2}$$

This system is obtained by setting $\sigma = 0$ on the right-hand side of the generic infeasible-interior-point equations (1.20). We now find the step lengths to the boundary along this direction, performing separate calculations for the primal and dual components as follows:

$$\begin{aligned}\alpha_{\rm aff}^{\rm pri} &= \arg\max\{\alpha \in [0,1] \,|\, x + \alpha\Delta x^{\rm aff} \geq 0\},\\ \alpha_{\rm aff}^{\rm dual} &= \arg\max\{\alpha \in [0,1] \,|\, s + \alpha\Delta s^{\rm aff} \geq 0\}.\end{aligned}$$

To measure the efficacy of the affine-scaling direction, we define $\mu_{\rm aff}$ as the hypothetical value of μ resulting from a full step to the boundary, that is,

$$\mu_{\rm aff} = (x + \alpha_{\rm aff}^{\rm pri}\Delta x^{\rm aff})^T(s + \alpha_{\rm aff}^{\rm dual}\Delta s^{\rm aff})/n.$$

If $\mu_{\rm aff} \ll \mu$, the affine-scaling direction is a good search direction that permits significant progress to be made in reducing μ, so we choose the centering parameter σ close to zero. If $\mu_{\rm aff}$ is just a little smaller than μ, we choose σ closer to 1. This choice has the effect of moving us closer to the central path $\mathcal{C}$, so that the algorithm is in a better position to achieve a substantial decrease in μ on the *next* iteration. Mehrotra [88] suggests the following heuristic, which has proved to be effective in exhaustive computational testing:

$$\sigma = \left(\frac{\mu_{\rm aff}}{\mu}\right)^3. \tag{10.3}$$

To calculate the centering step component, we could solve a linear system with the same coefficient matrix as in (10.1) and right-hand side $(0, 0, \sigma\mu e)$. As mentioned above, however, it is more efficient to calculate this term in conjunction with the corrector term, which we now describe.

To motivate the corrector step, let us see how the ith pairwise product $x_i s_i$ is affected by a full step in the affine-scaling direction. From (10.1), we have

$$\begin{aligned}&(x_i + \Delta x_i^{\rm aff})(s_i + \Delta s_i^{\rm aff}) \\ &= x_i s_i + x_i \Delta s_i^{\rm aff} + s_i \Delta x_i^{\rm aff} + \Delta x_i^{\rm aff}\Delta s_i^{\rm aff} = \Delta x_i^{\rm aff}\Delta s_i^{\rm aff}.\end{aligned} \tag{10.4}$$

When a full step is taken, the pairwise product $x_i s_i$ transforms to $\Delta x_i^{\rm aff}\Delta s_i^{\rm aff}$, instead of the zero value predicted by the linearized model used in (10.1). The corrector component $(\Delta x^{\rm cor}, \Delta\lambda^{\rm cor}, \Delta s^{\rm cor})$ tries to compensate for this deviation from linearity, modifying the search direction so that the pairwise products come closer to their target value of zero. This step satisfies the following linear system:

$$\begin{bmatrix} 0 & A^T & I \\ A & 0 & 0 \\ S & 0 & X \end{bmatrix} \begin{bmatrix} \Delta x^{\text{cor}} \\ \Delta \lambda^{\text{cor}} \\ \Delta s^{\text{cor}} \end{bmatrix} = \begin{bmatrix} 0 \\ 0 \\ -\Delta X^{\text{aff}} \Delta S^{\text{aff}} e \end{bmatrix}, \tag{10.5}$$

where

$$\begin{aligned} \Delta X^{\text{aff}} &= \text{diag}(\Delta x_1^{\text{aff}}, \Delta x_2^{\text{aff}}, \ldots, \Delta x_n^{\text{aff}}), \\ \Delta S^{\text{aff}} &= \text{diag}(\Delta s_1^{\text{aff}}, \Delta s_2^{\text{aff}}, \ldots, \Delta s_n^{\text{aff}}). \end{aligned}$$

To assess the effect of the corrector component, examine the pairwise product obtained from a full step along the combined affine-scaling/corrector direction. From (10.4) and (10.5) we obtain

$$\begin{aligned} & (x_i + \Delta x_i^{\text{aff}} + \Delta x_i^{\text{cor}})(s_i + \Delta s_i^{\text{aff}} + \Delta s_i^{\text{cor}}) \\ & \quad = \Delta x_i^{\text{aff}} \Delta s_i^{\text{cor}} + \Delta x_i^{\text{cor}} \Delta s_i^{\text{aff}} + \Delta x_i^{\text{cor}} \Delta s_i^{\text{cor}} \end{aligned} \tag{10.6}$$

(see the exercises). Is the pairwise product in (10.6) closer to zero than the one in (10.4)? When the coefficient matrix in (10.1) and (10.5) is approaching a nonsingular limit, the answer is yes. In this case, we have

$$\|(\Delta x^{\text{aff}}, \Delta s^{\text{aff}})\| = O(\mu), \qquad \|(\Delta x^{\text{cor}}, \Delta s^{\text{cor}})\| = O(\mu^2)$$

so that the right-hand side is $O(\mu^2)$ in (10.4) and $O(\mu^3)$ in (10.6). When the limiting matrix is singular, however, the corrector step may no longer be smaller in norm than the affine-scaling step—in fact, it is often larger. Even in this situation, however, use of the corrector component usually enhances the overall efficiency of the algorithm in practice.

The combined centering-corrector step $(\Delta x^{\text{cc}}, \Delta\lambda^{\text{cc}}, \Delta s^{\text{cc}})$ is obtained by solving the following linear system:

$$\begin{bmatrix} 0 & A^T & I \\ A & 0 & 0 \\ S & 0 & X \end{bmatrix} \begin{bmatrix} \Delta x^{\text{cc}} \\ \Delta \lambda^{\text{cc}} \\ \Delta s^{\text{cc}} \end{bmatrix} = \begin{bmatrix} 0 \\ 0 \\ \sigma\mu e - \Delta X^{\text{aff}} \Delta S^{\text{aff}} e \end{bmatrix}. \tag{10.7}$$

The factors of the coefficient matrix are already available from solving (10.1), so (10.7) can be solved at the cost of performing a single back-substitution.

Having described the essential elements of Mehrotra's approach, we specify a particular variant: Algorithm MPC.

Algorithm MPC
Given (x^0, λ^0, s^0) with $(x^0, s^0) > 0$;
for $k = 0, 1, 2, \ldots$
set $(x, \lambda, s) = (x^k, \lambda^k, s^k)$ and solve (10.1) for $(\Delta x^{\text{aff}}, \Delta\lambda^{\text{aff}}, \Delta s^{\text{aff}})$;
calculate

$$\begin{aligned} \alpha_{\text{aff}}^{\text{pri}} &= \arg\max\{\alpha \in [0,1] \,|\, x^k + \alpha \Delta x^{\text{aff}} \geq 0\}, &(10.8a)\\ \alpha_{\text{aff}}^{\text{dual}} &= \arg\max\{\alpha \in [0,1] \,|\, s^k + \alpha \Delta s^{\text{aff}} \geq 0\}, &(10.8b)\\ \mu_{\text{aff}} &= (x^k + \alpha_{\text{aff}}^{\text{pri}} \Delta x^{\text{aff}})^T (s^k + \alpha_{\text{aff}}^{\text{dual}} \Delta s^{\text{aff}})/n; &(10.8c) \end{aligned}$$

set centering parameter to $\sigma = (\mu_{\text{aff}}/\mu)^3$;
solve (10.7) for $(\Delta x^{\text{cc}}, \Delta\lambda^{\text{cc}}, \Delta s^{\text{cc}})$;
compute search direction and step to boundary from

$$\begin{aligned} (\Delta x^k, \Delta\lambda^k, \Delta s^k) &= (\Delta x^{\text{aff}}, \Delta\lambda^{\text{aff}}, \Delta s^{\text{aff}}) + (\Delta x^{\text{cc}}, \Delta\lambda^{\text{cc}}, \Delta s^{\text{cc}}); &(10.9a)\\ \alpha_{\max}^{\text{pri}} &= \arg\max\{\alpha \geq 0 \,|\, x^k + \alpha \Delta x^k \geq 0\}; &(10.9b)\\ \alpha_{\max}^{\text{dual}} &= \arg\max\{\alpha \geq 0 \,|\, s^k + \alpha \Delta s^k \geq 0\}; &(10.9c) \end{aligned}$$

set $\alpha_k^{\text{pri}} = \min(0.99 * \alpha_{\max}^{\text{pri}}, 1)$ and $\alpha_k^{\text{dual}} = \min(0.99 * \alpha_{\max}^{\text{dual}}, 1)$;
set

$$\begin{aligned} x^{k+1} &= x^k + \alpha_k^{\text{pri}} \Delta x^k, \\ (\lambda^{k+1}, s^{k+1}) &= (\lambda^k, s^k) + \alpha_k^{\text{dual}} (\Delta\lambda^k, \Delta s^k); \end{aligned}$$

end (for).

As defined above, Algorithm MPC is not quite the same as Mehrotra's original algorithm from [88]. However, it is similar to the variant that is implemented in codes such as LIPSOL, LOQO, and PCx (see Chapter 11).

Superquadratic Convergence

In the context of nonlinear equations, the idea of reusing the Jacobian to accelerate the convergence of Newton's method was proposed in the 1960s. Traub [132] stated an algorithm for scalar nonlinear equations and proved convergence at a superquadratic rate. Shamanskii [120] proved the same result for nonlinear systems.

We can state a variant of Shamanskii's approach as follows, for a general nonlinear mapping $F : \mathbb{R}^N \to \mathbb{R}^N$ whose Jacobian is denoted by $J(\cdot)$:

Algorithm S3 (to solve nonlinear equations $F(z) = 0$)
Choose $z^0 \in \mathbb{R}^N$;
for $k = 0, 1, 2, \ldots$
  compute $d^k = -J(z^k)^{-1}F(z^k)$ and choose step length α_k;
  $z \leftarrow z^k + \alpha_k d^k$;
  compute $d = -J(z^k)^{-1}F(z)$ and choose step length α;
  if $\|F(z + \alpha d)\| \leq 0.9 * \|F(z)\|$
    $z \leftarrow z + \alpha d$;
  end (if)
  $z^{k+1} \leftarrow z$;
end(for).

At each iteration, this method takes a step along the usual Newton direction d^k. It then computes a Newton-like step d using the new function vector $F(z)$ but the old Jacobian $J(z^k)$. If a step along d produces a reasonable reduction in $\|F\|$, we take it; otherwise, we ignore it. The step d costs less to compute than d^k, since no reevaluation and refactorization of the Jacobian $J(z^k)$ are needed. The step lengths α_k and α are chosen by standard line search techniques.

When Algorithm S3 approaches a nondegenerate solution z^* (at which $F(z^*) = 0$ and $J(z^*)$ is nonsingular), it converges cubically, provided that the step lengths α_k and α both approach 1 rapidly enough.

The direction d^k in Algorithm S3 obviously corresponds to the affine-scaling direction (10.1) in Algorithm MPC. However, the corrector direction in Algorithm MPC corresponds to d in Algorithm S3 only if we take a full step along $(\Delta x^{\text{aff}}, \Delta \lambda^{\text{aff}}, \Delta s^{\text{aff}})$ and eliminate centering by setting $\sigma = 0$. Neither condition is satisfied in practice, so the correspondence between the two methods is imperfect.

Wright and Zhang [151] propose an infeasible primal-dual algorithm that more closely follows Algorithm S3, modified to keep the iterates in a neighborhood of the form $\mathcal{N}_{-\infty}(\gamma, \beta)$ (6.2) and to allow multiple corrections. When the maximum number of corrections allowed on each iteration is fixed at I, the asymptotic convergence has Q-order $I + 2$. Surprisingly, this result holds even when there are multiple solutions, that is, when the coefficient matrix in (10.1) approaches a singular limit.

Second-Order Trajectory-Following Methods

Trajectory-following methods from ODEs give a slightly different (but related) perspective on Mehrotra's algorithm. These methods define a trajectory from the current point (x, λ, s) to the solution set Ω and then follow this trajectory along its length to obtain a solution. There are infinitely many trajectories to choose from, in general. The most obvious one in our case is obtained from a linear scaling of the function F defined in (1.5). Denoting this trajectory by $\mathcal{H}$ and parametrizing it by $\tau \in [0, 1)$, we find that each point $(\hat{x}_\tau, \hat{\lambda}_\tau, \hat{s}_\tau) \in \mathcal{H}$ is a solution of the following nonlinear system:

$$\begin{bmatrix} A^T\hat{\lambda} + \hat{s} - c \\ A\hat{x} - b \\ \hat{X}\hat{S}e \end{bmatrix} = \begin{bmatrix} (1-\tau)r_c \\ (1-\tau)r_b \\ (1-\tau)XSe \end{bmatrix}, \qquad (\hat{x}, \hat{s}) \geq 0. \tag{10.10}$$

It is easy to see that $(\hat{x}_0, \hat{\lambda}_0, \hat{s}_0) = (x, \lambda, s)$ and that, if the limit of $(\hat{x}_\tau, \hat{\lambda}_\tau, \hat{s}_\tau)$ exists as $\tau \uparrow 1$, this limit point is a primal-dual solution. The trajectory $\mathcal{H}$ was illustrated in Figure 1.1.

To move along $\mathcal{H}$, we can form a Taylor series approximation to $(\hat{x}_\tau, \hat{\lambda}_\tau, \hat{s}_\tau)$ by expanding about the initial point (x, λ, s) as follows:

$$\begin{aligned} &(\hat{x}_\tau, \hat{\lambda}_\tau, \hat{s}_\tau) \\ &\quad = (\hat{x}_0, \hat{\lambda}_0, \hat{s}_0) + \tau(\hat{x}_0', \hat{\lambda}_0', \hat{s}_0') + \tfrac{1}{2}\tau^2(\hat{x}_0'', \hat{\lambda}_0'', \hat{s}_0'') + \cdots \\ &\quad = \sum_{j=0}^{\infty} \frac{\tau^j}{j!}(\hat{x}_0^{(j)}, \hat{\lambda}_0^{(j)}, \hat{s}_0^{(j)}). \end{aligned} \tag{10.11}$$

Here, $(\hat{x}_0^{(j)}, \hat{\lambda}_0^{(j)}, \hat{s}_0^{(j)})$ denotes the jth derivative of $(\hat{x}_\tau, \hat{\lambda}_\tau, \hat{s}_\tau)$ with respect to the parameter τ, evaluated at $\tau = 0$. We can find these derivatives by implicitly differentiating both sides of the equality condition in (10.10). For the first derivative—geometrically, the tangent to $\mathcal{H}$—we have

$$\begin{bmatrix} 0 & A^T & I \\ A & 0 & 0 \\ \hat{S}_\tau & 0 & \hat{X}_\tau \end{bmatrix} \begin{bmatrix} \hat{x}_\tau' \\ \hat{\lambda}_\tau' \\ \hat{s}_\tau' \end{bmatrix} = \begin{bmatrix} -r_c \\ -r_b \\ -XSe \end{bmatrix}. \tag{10.12}$$

Setting $\tau = 0$ and noting that $\hat{x}_0 = x$ and $\hat{s}_0 = s$, we have

$$\begin{bmatrix} 0 & A^T & I \\ A & 0 & 0 \\ S & 0 & X \end{bmatrix} \begin{bmatrix} \hat{x}_0' \\ \hat{\lambda}_0' \\ \hat{s}_0' \end{bmatrix} = \begin{bmatrix} -r_c \\ -r_b \\ -XSe \end{bmatrix},$$

which is exactly the same system as the affine-scaling step equation (10.1). Hence, we can make the identification

$$(\Delta x^{\text{aff}}, \Delta\lambda^{\text{aff}}, \Delta s^{\text{aff}}) = (\hat{x}_0', \hat{\lambda}_0', \hat{s}_0'). \tag{10.13}$$

Differentiating (10.12) again with respect to τ, we find that the second derivative satisfies the following system:

$$\begin{bmatrix} 0 & A^T & I \\ A & 0 & 0 \\ \hat{S}_\tau & 0 & \hat{X}_\tau \end{bmatrix} \begin{bmatrix} \hat{x}''_\tau \\ \hat{\lambda}''_\tau \\ \hat{s}''_\tau \end{bmatrix} = \begin{bmatrix} 0 \\ 0 \\ -2\hat{X}'_\tau \hat{S}'_\tau e \end{bmatrix}. \tag{10.14}$$

Substituting $\tau = 0$ into (10.14), noting (10.13), and comparing with (10.5), we see that

$$(\Delta x^{\text{cor}}, \Delta\lambda^{\text{cor}}, \Delta s^{\text{cor}}) = \tfrac{1}{2}(\hat{x}''_0, \hat{\lambda}''_0, \hat{s}''_0). \tag{10.15}$$

If we truncate the Taylor series in (10.11) at two terms and use the identities (10.13) and (10.15), we obtain the following quadratic approximation to the trajectory $\mathcal{H}$:

$$\begin{aligned} &(\hat{x}_\tau, \hat{\lambda}_\tau, \hat{s}_\tau) \\ &\approx \; (x, \lambda, s) + \tau(\Delta x^{\text{aff}}, \Delta\lambda^{\text{aff}}, \Delta s^{\text{aff}}) + \tau^2(\Delta x^{\text{cor}}, \Delta\lambda^{\text{cor}}, \Delta s^{\text{cor}}). \end{aligned} \tag{10.16}$$

How does this approximation compare with the line search used in Algorithm MPC? If we ignore the centering term by setting $\sigma = 0$ and constrain the primal and dual step lengths to be identical, we see from (10.9) that Algorithm MPC searches along the line

$$(x, \lambda, s) + \tau(\Delta x^{\text{aff}}, \Delta\lambda^{\text{aff}}, \Delta s^{\text{aff}}) + \tau(\Delta x^{\text{cor}}, \Delta\lambda^{\text{cor}}, \Delta s^{\text{cor}}); \tag{10.17}$$

that is, the τ^2 coefficient in the last term is replaced by τ. Figure 10.1 shows how the curved approximation (10.16) differs from the linear approximation (10.17). The difference may be significant, since μ always decreases for τ small and positive along the path defined by (10.16), whereas it may actually increase along the path defined by (10.17).

When we account for the centering component, this simple correspondence between the derivatives of the trajectory and the components of the search direction used by Algorithm MPC no longer holds. Mehrotra [87] defines a modified trajectory $\mathcal{H}_\sigma$ for which the correspondence continues to hold even in the presence of centering. Like $\mathcal{H}$, the trajectory $\mathcal{H}_\sigma$ starts at the current point (x, λ, s) and aims at the solution set Ω. The general point $(\hat{x}_\tau, \hat{\lambda}_\tau, \hat{s}_\tau) \in \mathcal{H}_\sigma$ satisfies the following system:

$$\begin{bmatrix} A^T\hat{\lambda} + \hat{s} - c \\ A\hat{x} - b \\ \hat{X}\hat{S}e \end{bmatrix} = \begin{bmatrix} (1-\tau)r_c \\ (1-\tau)r_b \\ (1-\tau)XSe + \tau^2(1-\tau)\sigma\mu e \end{bmatrix}. \tag{10.18}$$

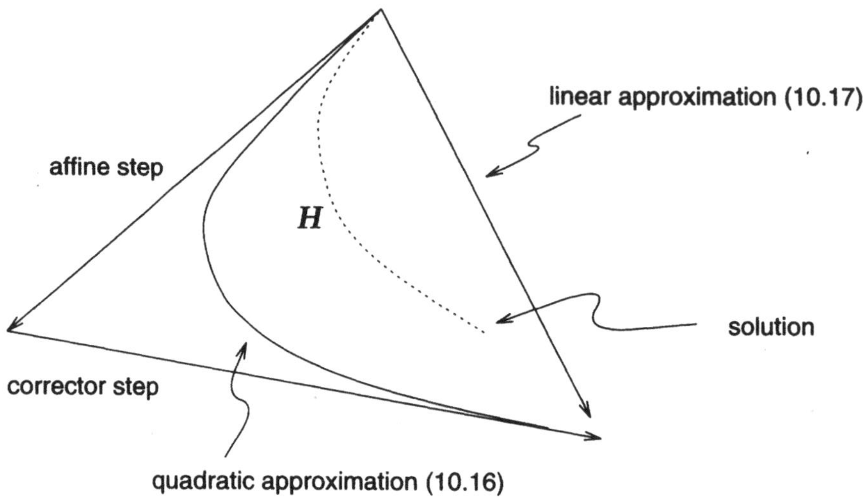

Figure 10.1. Curved and linear approximations to $\mathcal{H}$.

The tangent to this trajectory is the affine-scaling direction, as for $\mathcal{H}$, but the curvature can be identified with the combined centering-corrector step rather than the corrector component alone. That is, we have

$$(\Delta x^{\text{aff}}, \Delta\lambda^{\text{aff}}, \Delta s^{\text{aff}}) = (\hat{x}'_0, \hat{\lambda}'_0, \hat{s}'_0), \tag{10.19a}$$

$$(\Delta x^{\text{cc}}, \Delta\lambda^{\text{cc}}, \Delta s^{\text{cc}}) = \tfrac{1}{2}(\hat{x}''_0, \hat{\lambda}''_0, \hat{s}''_0) \tag{10.19b}$$

(see the exercises). The trajectory $\mathcal{H}_\sigma$ tends to bulge more toward the central path than does $\mathcal{H}$, as we show in Figure 10.2.

Despite the closer relationship between $\mathcal{H}_\sigma$ and Algorithm MPC, we cannot quite classify the algorithm as a second-order trajectory-following algorithm because it still searches along a linear direction rather than a quadratic path.

Higher-Order Methods

We have already noted that Mehrotra's second-order algorithm tends to perform better than the algorithms from Framework PD, which rely on first-order approximations of paths to the solution set Ω. This observation leads us to ask if we can do even better by taking a few more terms from the Taylor series to form third- and fourth-order approximations to the trajectories $\mathcal{H}$ and $\mathcal{H}_\sigma$. Each higher derivative $(\hat{x}_\tau^{(j)}, \hat{\lambda}_\tau^{(j)}, \hat{s}_\tau^{(j)})$ can be found for the cost of a substitution, together with a few componentwise vector products. Do the advantages of a higher-order approximation outweigh the extra cost?

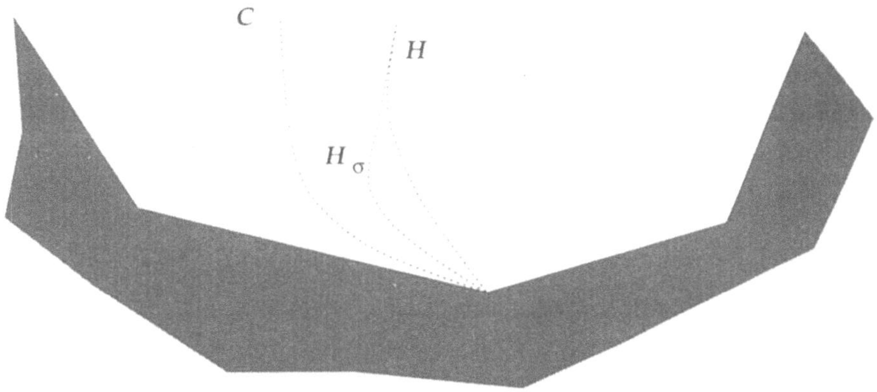

Figure 10.2. Central path $\mathcal{C}$ and trajectories $\mathcal{H}$ and $\mathcal{H}_\sigma$ from (x, λ, s) to the solution set Ω.

The answer to this question is not yet clear. Carpenter et al. [20] experimented with multiple-correction versions of Mehrotra's algorithm and concluded that the original strategy of a single correction is optimal. They found that additional corrections generally reduce the iteration count but not the computation time. Computational experience like this has confirmed that the effectiveness of methods based on Taylor series is limited by the awkward geometry of the trajectories $\mathcal{C}$, $\mathcal{H}$, and $\mathcal{H}_\sigma$. The central path $\mathcal{C}$ has been described as a number of smooth segments connected by sharp turns (Vavasis and Ye [141]). At the turns, the higher-order derivatives with respect to the arc length parameter τ are large, so we need many terms in the Taylor summation to obtain a reasonable approximation (see Figure 10.3). Path-following algorithms invariably find it difficult to negotiate the turns, and third- or fourth-order approximations do not fare much better than the first- and second-order approximations.

Gondzio [43] recognizes the difficulty of following the trajectory, and he proposes a slightly different use of higher-order information. His algorithm computes the usual Mehrotra search direction and enhances it with corrector components of successively higher order in an attempt to increase the step length α. Rather than targeting a point on the central path, however, the additional corrector steps aim only to move back into a loose infeasible neighborhood of $\mathcal{C}$. This less ambitious goal is achieved well enough that Gondzio reports computational savings of around 20% over Mehrotra's algorithm on a standard test set. Gondzio's algorithm is implemented in the code HOPDM, and his modifications are being incorporated gradually into

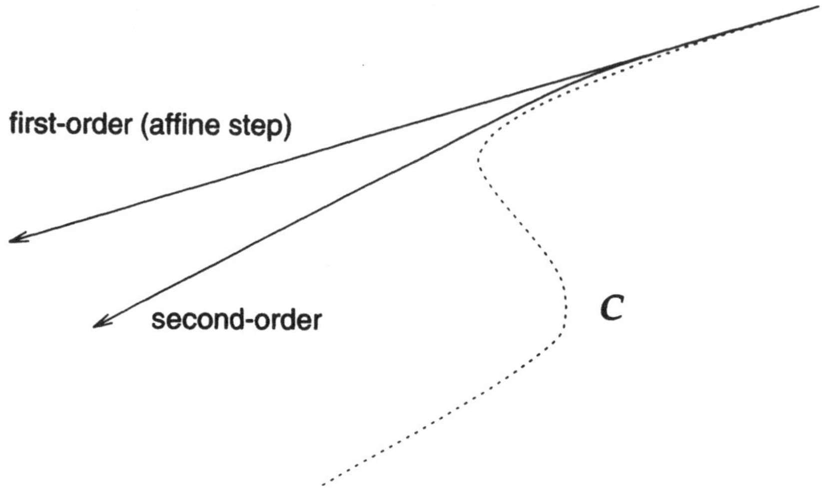

Figure 10.3. Low-order Taylor series approximations to $\mathcal{C}$ have trouble negotiating its numerous sharp turns.

other codes as well (see Appendix B).

The geometry of $\mathcal{C}$ is related to the data-dependent quantity $\epsilon(A,b,c)$ defined in (3.5). As we show in Chapter 7, the index set $\{1,2,\ldots,n\}$ is resolved into its $\mathcal{B}\cup\mathcal{N}$ partition once μ reaches the level of $\epsilon(A,b,c)^2$. At this point, fast asymptotic convergence and finite termination properties take hold. Geometrically, $\epsilon(A,b,c)^2$ defines a point on the central path $\mathcal{C}$ beyond which there are no more sharp turns. For τ from (1.9) in the approximate range $(0,\epsilon(A,b,c)^2)$, the central path flattens out for its final approach to the solution set Ω.

Further Enhancements

We mention a few more enhancements of Mehrotra's basic algorithm that improve its efficiency and robustness in certain situations. These include an adaptive choice of step length, a modification of the choice of centering parameter σ, and a variation on the search direction that was described in Mehrotra's original paper.

In Algorithm MPC, we apply the scaling factor .99 to the step-to-boundary values $\alpha_{\max}^{\text{pri}}$ and $\alpha_{\max}^{\text{dual}}$ defined in (10.9b), (10.9c). Other codes use different multiples, ranging from conservative values such as .9 or .95 to the aggressive .9999. The code LIPSOL [162] uses an adaptive strategy to prevent the pairwise products x_is_i from becoming too unbalanced. It first tries a multiplier very close to 1 and checks that the new iterate stays inside a neighborhood like $\mathcal{N}_{-\infty}(\gamma)$ for some small value of γ. If not, it backs off along the chosen direction, trying smaller and smaller values for the step length until the neighborhood condition is satisfied.

Mehrotra [88, section 6] defines a heuristic in which this scaling factor applied to $\alpha_k^{\rm pri}$ and $\alpha_k^{\rm dual}$ approaches 1 on later iterations. He chooses this factor to be the largest value for which the "critical" pairwise products $x_i s_i$—the ones that determine the values of $\alpha_{\rm max}^{\rm pri}$ and $\alpha_{\rm max}^{\rm dual}$ in (10.9b) and (10.9c)—are larger than $\gamma_f \mu_+$, where γ_f is a small constant and μ_+ is an approximation to μ_{k+1}. The procedure can be outlined formally as follows:

Given a parameter γ_f with $0 < \gamma_f \ll 1$, and
$(\Delta x^k, \Delta\lambda^k, \Delta s^k)$, $\alpha_{\rm max}^{\rm pri}$, and $\alpha_{\rm max}^{\rm dual}$ from (10.9);
calculate

$$\mu_+ = (x^k + \alpha_{\rm max}^{\rm pri}\Delta x^k)^T(s^k + \alpha_{\rm max}^{\rm dual}\Delta s^k)/n;$$

for the particular index i for which $x_i^k + \alpha_{\rm max}^{\rm pri}\Delta x_i^k = 0$, define $f^{\rm pri}$ by

$$(x_i^k + f^{\rm pri}\alpha_{\rm max}^{\rm pri}\Delta x_i^k)(s_i^k + \alpha_{\rm max}^{\rm dual}\Delta s_i^k) = \gamma_f \mu_+;$$

for the particular index i for which $s_i^k + \alpha_{\rm max}^{\rm dual}\Delta s_i^k = 0$, define $f^{\rm dual}$ by

$$(x_i^k + \alpha_{\rm max}^{\rm pri}\Delta x_i^k)(s_i^k + f^{\rm dual}\alpha_{\rm max}^{\rm dual}\Delta s_i^k) = \gamma_f \mu_+;$$

set

$$\alpha_k^{\rm pri} = \max(1-\gamma_f, f^{\rm pri})\alpha_{\rm max}^{\rm pri}, \qquad \alpha_k^{\rm dual} = \max(1-\gamma_f, f^{\rm dual})\alpha_{\rm max}^{\rm dual}.$$

A typical value for the parameter γ_f is .01. This heuristic speeds the asymptotic convergence and also improves the robustness of the algorithm. Testing on standard problems shows that, in a number of cases, the use of this heuristic produces a solution, while the use of a step factor bounded away from 1 causes failure.

One minor variant is in the choice of centering parameter σ in (10.7). Mehrotra actually uses a fairly complex procedure for choosing this parameter in his original paper [88]. For one thing, he experiments with the exponent in (10.3), comparing the values 1, 2, 3, and 4 with the no-centering

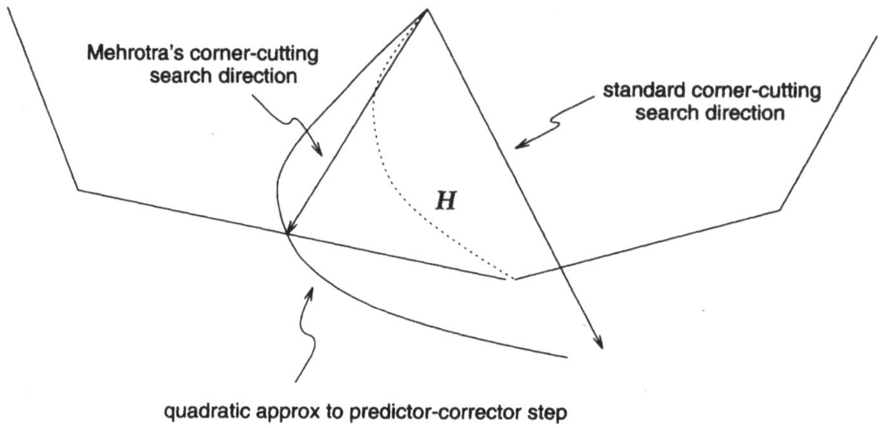

Figure 10.4. Comparison of Mehrotra's search direction, standard corner-cutting search direction, and quadratic interpolant.

choice $\sigma = 0$. The exponents 2, 3, and 4 all give reasonable performance on his limited test set. In later testing, Lustig, Marsten, and Shanno [78] favored an exponent of 2, although they too noted that the results were not too sensitive to this value. Mehrotra modifies σ further by looking at the effect of the residuals r_b and r_c on the step lengths along the affine-scaling direction $\alpha_{\text{aff}}^{\text{pri}}$ and $\alpha_{\text{aff}}^{\text{dual}}$, defined in (10.8a) and (10.8b). If these residuals are nonzero and if they cause reductions in $\alpha_{\text{aff}}^{\text{pri}}$ and $\alpha_{\text{aff}}^{\text{dual}}$, he increases σ to place more emphasis on centering. He notes that this strategy apparently helps the algorithm to avoid troublesome regions of slow progress.

The original algorithm of Mehrotra actually uses a slightly different search direction from Algorithm MPC. Instead of adding the affine-scaling direction to the centering/corrector direction as in (10.9a), Mehrotra defines the parameters ϵ_x and ϵ_s as the maximum steps to the boundary along a curved path:

$$\begin{aligned}\epsilon_x &= \arg\max\{\epsilon \in [0,1] \,|\, x + \epsilon\Delta x^{\text{aff}} + \epsilon^2\Delta x^{\text{cc}} \geq 0\},\\ \epsilon_s &= \arg\max\{\epsilon \in [0,1] \,|\, s + \epsilon\Delta s^{\text{aff}} + \epsilon^2\Delta s^{\text{cc}} \geq 0\}.\end{aligned}$$

He then chooses the search direction to be

$$(\Delta x, \Delta\lambda, \Delta s) = (\epsilon_x\Delta x^{\text{aff}}, \epsilon_s\Delta\lambda^{\text{aff}}, \epsilon_s\Delta s^{\text{aff}}) + (\epsilon_x^2\Delta x^{\text{cc}}, \epsilon_s^2\Delta\lambda^{\text{cc}}, \epsilon_s^2\Delta s^{\text{cc}}) \tag{10.20}$$

(see Figure 10.4). As we discussed earlier, the path (10.20) is a second-order approximation of the trajectory $\mathcal{H}_\sigma$. Mehrotra's strategy still "cuts the corner" of the quadratic interpolant, albeit in a slightly different way from Algorithm MPC, so the relationship with second-order trajectory following remains imprecise.

Notes and References

A cautionary note: The development of primal-dual codes has been heavily influenced by the Netlib collection of linear programming test problems, which was set up during the 1980s for this very purpose. Although the approach we have described in this chapter seems to be the winner on this particular set of test problems, it is possible that future research will reveal methods that perform even better. In any case, it is not wise to let the Netlib test set be the final arbiter in deciding issues of practical performance. Typical users of linear programming software often need to solve just a few specific types of problems, and it may happen that some method other than the one described here is more efficient for their particular applications.

Exercises

1. Verify equation (10.6).

2. Let the affine-scaling and corrector directions be defined as in (10.1) and (10.5), respectively, and consider a step of the form

$$(x, \lambda, s) + \alpha_p(\Delta x^{\text{aff}}, \Delta\lambda^{\text{aff}}, \Delta s^{\text{aff}}) + \alpha_c(\Delta x^{\text{cor}}, \Delta\lambda^{\text{cor}}, \Delta s^{\text{cor}}).$$

   Compute the value of $x_i s_i$ as a function of α_p and α_c along this direction. What relationship between the two step parameters ensures that no cross term $\Delta x_i^{\text{aff}} \Delta s_i^{\text{aff}}$ appears in this expression?

3. Taking (10.12) and (10.14) as a starting point, find a general formula for the derivative $(\hat{x}_\tau^{(j)}, \hat{\lambda}_\tau^{(j)}, \hat{s}_\tau^{(j)})$ along the trajectory $\mathcal{H}$, for arbitrary j.

4. Verify that the relationships (10.19) hold for the first and second derivatives of the trajectory $\mathcal{H}_\sigma$ defined by (10.18).

5. Suppose that the step $(\Delta x, \Delta\lambda, \Delta s)$ satisfies (1.20) and different step lengths α^{pri} and α^{dual} are used for the primal and dual variables. Show how the duality gap $x^T s$ varies with α^{pri} and α^{dual} as we move along this direction.

6. Consider the generalization of the Newton-like step equation (1.13) to the convex QP problem (8.11), as derived in the exercises for Chapter 8. If different step lengths $\alpha^{\rm pri}$ and $\alpha^{\rm dual}$ are used for the primal and dual variables, show that the new iterate is not necessarily feasible with respect to the KKT condition $Qx + A^T\lambda + s = c$.

Chapter 11

Implementations

Our major concern in this book has been the theory of primal-dual algorithms, although we have been guided and motivated throughout by issues of practical performance. Of course, a successful piece of optimization software requires more than just a good optimization algorithm. Efficiency and stability of the linear algebra are important ingredients, as they are for most other types of numerical software. Input and output should be straightforward (or, at least, should adhere to the standards of the field). The code should initialize and terminate the algorithm in a reasonable way and choose suitable values for the algorithmic parameters while allowing expert users to override its default settings. If in the public domain, the code should be written well enough to allow users to reuse components or to experiment with variations of the algorithm.

In this chapter, we look at some of these nitty-gritty issues as they relate to primal-dual codes. The most important issue is the linear algebra—the problem of solving a large, sparse linear system at each iteration to find the search direction. The sparse linear algebra module is by far the most complex component of a typical primal-dual code. Fortunately, numerical analysts have done most of the work in devising efficient algorithms for such systems, and only minor modifications are needed to make their algorithms fit our purpose. Other issues that we touch on here include initialization of the algorithm, termination criteria, preprocessing of the data, and handling of free variables. Our treatment is far from comprehensive or exhaustive; we aim to convey the flavor of each topic, outlining the basic issues and providing some pointers to the literature.

Appendix B contains brief summaries of the current roster of widely available software. This information is subject to change, so we plan to maintain a World Wide Web page with up-to-date information at the following URL:

```
http://www.siam.org/books/swright/
```

Three Forms of the Step Equation

We saw in Chapter 1 that the linear systems to be solved at each primal-dual iteration can be formulated in three equivalent ways. We restate them here, changing the notation slightly for generality. The unreduced form is

$$\begin{bmatrix} 0 & A & 0 \\ A^T & 0 & I \\ 0 & S & X \end{bmatrix} \begin{bmatrix} \Delta\lambda \\ \Delta x \\ \Delta s \end{bmatrix} = \begin{bmatrix} -r_b \\ -r_c \\ -r_{xs} \end{bmatrix}, \tag{11.1}$$

where $r_{xs} = XSe - \sigma\mu e$ for the generic primal-dual step (see (1.12) and (1.20)), whereas the predictor and corrector steps in Mehrotra's algorithm use $r_{xs} = XSe$ and $r_{xs} = -\sigma\mu e + \Delta X^{\text{aff}} \Delta S^{\text{aff}} e$, respectively (see (10.1) and (10.7)). Eliminating Δs from (11.1) and using the notation $D = S^{-1/2}X^{1/2}$ from (1.24), we obtain the following form, often known as the *augmented system* form:

$$\begin{aligned} \begin{bmatrix} 0 & A \\ A^T & -D^{-2} \end{bmatrix} \begin{bmatrix} \Delta\lambda \\ \Delta x \end{bmatrix} &= \begin{bmatrix} -r_b \\ -r_c + X^{-1} r_{xs} \end{bmatrix}, & (11.2a) \\ \Delta s &= -X^{-1}(r_{xs} + S\Delta x). & (11.2b) \end{aligned}$$

Finally, we can eliminate Δx from (11.2a) to obtain the most compact of the three forms:

$$\begin{aligned} AD^2A^T\Delta\lambda &= -r_b + A(-S^{-1}Xr_c + S^{-1}r_{xs}), & (11.3a) \\ \Delta s &= -r_c - A^T\Delta\lambda, & (11.3b) \\ \Delta x &= -S^{-1}(r_{xs} + X\Delta s). & (11.3c) \end{aligned}$$

This form is sometimes known as the *normal equations* form because, in the case of $r_b = 0$, the equations (11.3a) are the normal equations for a linear least squares problem with coefficient matrix DA^T.

The normal equations form is used by most primal-dual codes because the symmetric positive semidefinite coefficient matrix in (11.3a) can be factored by sparse Cholesky techniques, for which algorithms and software are readily available. The coefficient matrix in (11.2a) is symmetric but indefinite, and algorithms for factoring it are more complicated. The augmented system has advantages in stability and flexibility, however. It has been used in some experimental codes and will probably appear in "production" codes in the near future. For both (11.2) and (11.3), particular issues of stability arise because of possible rank deficiency in A and the presence of very small and very large diagonal elements in both D and D^{-1}. We take special note of these issues in our discussion of algorithms below.

The Cholesky Factorization

Let M denote the symmetric coefficient matrix in (11.3a) after a possible symmetric reordering of its rows and columns; that is,

$$M = P(AD^2A^T)P^T, \tag{11.4}$$

where P is an $m \times m$ permutation matrix. Assume for the moment that A has full row rank so that M is positive definite. The Cholesky algorithm produces a lower triangular matrix L with positive diagonal elements such that

$$LL^T = M.$$

Moreover, L is unique. The algorithm is conceptually quite simple: one variant is completely described by the following pseudocode.

for $i = 1$ **to** m
    $L_{ii} \leftarrow \sqrt{M_{ii}}$;
    for $j = i + 1$ **to** m
        $L_{ji} \leftarrow M_{ji}/L_{ii}$;
        for $k = i + 1$ **to** j
            $M_{jk} \leftarrow M_{jk} - L_{ji}L_{ki}$;
        end (for)
    end (for)
end (for).

Algebraically, we can write the update step (the k loop) as

$$M_{i+1:m,i+1:m} \leftarrow M_{i+1:m,i+1:m} - M_{i+1:m,i}M^T_{i+1:m,i}/M_{ii}. \tag{11.5}$$

The element M_{ii} is called the *pivot element.*

Even if we neglect the sparsity of M, there are many ways to reorganize the computations in the Cholesky factorization to suit different computer architectures. For instance, the updating of each remaining column of M

in the k loop can be delayed until just before that column is converted to a column of L. Also, the columns can be processed in groups so that the vector-vector arithmetic in the innermost loop becomes matrix-vector or even matrix-matrix arithmetic, making it more suitable for high-performance computers with cache memories.

Given the factorization $M = LL^T$, the linear system $Mz = r$ can be solved with two triangular substitutions:

$$\begin{aligned} &\text{solve } Ly = r \text{ to find } y; \\ &\text{solve } L^T z = y \text{ to find } z. \end{aligned}$$

The process of Cholesky factorization followed by triangular substitutions is a remarkably stable way to compute solutions of linear systems with symmetric positive definite coefficient matrices. If we denote the unit roundoff by $\mathbf{u}$ (typically, $\mathbf{u} \approx 10^{-16}$ for double-precision arithmetic), the inexact solution $\tilde{z}$ obtained from an implementation of this algorithm is related to the exact solution z by the following expression:

$$\frac{\|\tilde{z} - z\|}{\|z\|} \leq \gamma\kappa(M)\mathbf{u}, \tag{11.6}$$

where $\kappa(M)$ denotes the condition number of M. Note that the estimate (11.6) holds *independently of the permutation matrix P in* (11.4), that is, independently of the row and column ordering. This fact is particularly significant in dealing with sparse matrices M. In this case, we can choose the permutation P to ensure that the factor L is sparse, without worrying about the effect of P on the accuracy of the solution vector.

Sparse Cholesky Factorization: Minimum-Degree Orderings

Cholesky factorization of a sparse matrix usually produces *fill-in*; that is, some lower triangular locations in the factor L contain nonzero elements even though the same locations in the original matrix M are zero. This process can be illustrated with the simple example of a 5×5 matrix whose nonzero pattern is depicted on the left of the following diagram:

$$\begin{bmatrix} \times & & & & \\ 0 & \times & & & \\ \times & \times & \times & & \\ \times & 0 & 0 & \times & \\ 0 & 0 & 0 & \times & \times \end{bmatrix} \rightarrow \left[\begin{array}{c|cccc} \times & & & & \\ \cline{2-5} 0 & \times & & & \\ \times & \times & \times & & \\ \times & 0 & + & \times & \\ 0 & 0 & 0 & \times & \times \end{array}\right]. \tag{11.7}$$

(We show only the lower triangle of the matrix—the upper triangle is simply its transpose, because of symmetry—and denote each nonzero element by an "×.") Suppose that we perform one step of Cholesky factorization on this matrix, using the $(1, 1)$ element as the pivot element. During this step, the lower left 4×4 submatrix is updated by a weighted outer product of the first column with itself, leading to fill-in in the $(4, 3)$ location, depicted by a "+" in the right-hand portion of (11.7). The updated lower left 4×4 submatrix is called the *remaining matrix*. The factorization proceeds recursively by applying another elimination step to the remaining matrix at each step until only a null matrix remains.

We can exercise some control over the amount of fill-in through our choice of the row/column ordering in M, that is, our choice of the permutation matrix P. We would like to keep fill-in small for two reasons. First, fill-in requires additional storage. (See Appendix A for details on sparse matrix storage.) Second, the extra nonzeros add to the cost of updating the remaining matrix (the k loop in the algorithm above) and also to the cost of the triangular substitutions with L and L^T that are needed to recover the solution vector.

The problem of finding the ordering that actually *minimizes* fill-in is too difficult to solve in general; in fact, it is NP-complete [117]. Fortunately, there are inexpensive ordering heuristics that perform quite well on most practical examples—the amounts of fill-in resulting from these heuristics are not too much greater than the minimum.

Possibly the best-known class of ordering heuristics goes by the name of *minimum degree*. The term *degree* refers to the number of nonzero elements in a column of the matrix, excluding the diagonal. At each step, this heuristic examines the remaining matrix and selects the diagonal element with minimum degree—the one with fewest nonzeros in its row/column. If this element occurs in the (ℓ, ℓ) position, we swap rows 1 and ℓ and columns 1 and ℓ of the remaining matrix. We then perform the elimination as in (11.7), using the newly established $(1, 1)$ element as the pivot element.

Minimum-degree orderings are motivated by the formula (11.5): If the pivot element $M_{\ell\ell}$ has degree d, the update matrix contains d^2 nonzeros. Therefore, it makes sense to choose the pivot so that d is as small as possible.

The five columns of the left-hand matrix in (11.7) have 3, 2, 3, 3, and 2 nonzeros, respectively, so their degrees are 2, 1, 2, 2, and 1, respectively. (Remember, only the lower triangle is shown in (11.7); we have to count the nonzeros in the upper triangle as well.) The minimum-degree heuristic would choose either the $(2, 2)$ or the $(5, 5)$ element as the pivot, not the $(1, 1)$

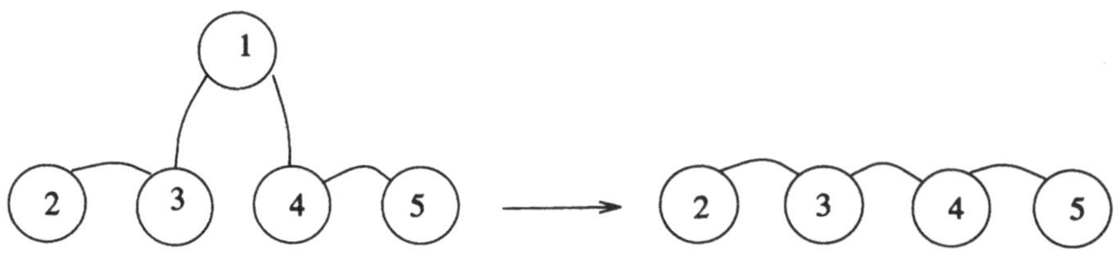

Figure 11.1. Adjacency graphs for the matrices in (11.7), showing the effect of elimination.

element as we did in (11.7). The diagonals of the remaining matrix on the right-hand side of (11.7) have degrees 1, 2, 2, and 1, respectively.

In sparse Cholesky codes, ordering and factorization do not take place concurrently, as we suggested in the discussion above. Rather, the ordering is performed prior to any numerical operations by making use of an *adjacency graph*, which is a collection of nodes and arcs that depicts the locations of the nonzeros in the original matrix and all the remaining matrices. This graph consists initially of m nodes, with an edge connecting nodes i and j whenever the (i, j) element of the matrix is nonzero. Nodes i and j are called *neighbors* or *adjacent nodes*, and the edge that joins them is said to be *incident* to both nodes. The *degree* of node i is simply the number of edges incident to node i.

The effect of eliminating row/column i during a step of Cholesky is captured by an elimination operation on the graph: If we discard node i and then create edges between all the nodes that were previously connected to node i, except where such edges exist already, the remaining graph represents the nonzero structure of the remaining matrix that results from this elimination step. The fill-in at this step equals the number of new edges that were created. Figure 11.1 illustrates the node elimination procedure for the example (11.7).

Many variants of minimum degree have been proposed; see George and Liu [39] for a summary. All these variants try to reduce the overhead associated with keeping track of the adjacency graph as it evolves during the factorization. The most important variants use the concept of *supernodes*, where a supernode is defined as a maximal set of nodes Θ for which

(i) all nodes in Θ are neighbors of one another;

(ii) each node in Θ has the same set of neighbors from outside Θ.

In terms of the matrix, each supernode corresponds to a maximal group of columns with identical nonzero patterns. For the following 5×5 matrix, for example, the node set $\{1, 3\}$ is a supernode, because columns 1 and 3 both have nonzeros in positions 1, 3, and 4:

$$\begin{bmatrix} \times & & & & \\ 0 & \times & & & \\ \times & 0 & \times & & \\ \times & \times & \times & \times & \\ 0 & \times & 0 & \times & \times \end{bmatrix}. \tag{11.8}$$

It follows immediately from the definition that all nodes in a supernode have the same degree. Supernodes can be created in the remaining graph during the factorization process, and they cannot be broken up by subsequent elimination steps. Exploiting these properties of supernodes, we can reduce the ordering time. The adjacency graph can be stored in a more compact form and needs to be updated less frequently.

If the minimum-degree heuristic selects one node from a supernode Θ for elimination, the remaining nodes in Θ still have minimum degree in the remaining graph that results from the elimination step. Hence, all nodes in a supernode are selected consecutively for elimination. We can exploit this fact in performing numerical elimination on the matrix. Arithmetic can be performed jointly on all the columns of a supernode, thus reducing the need for packing and unpacking of sparse vectors and increasing the amount of dense matrix-vector arithmetic.

Other Orderings

Although minimum-degree orderings remain the most popular orderings in primal-dual codes that use the normal equations formulation (11.3) (and indeed in all implementation of sparse Cholesky for general matrices), some recent experiments have focused on other ordering heuristics.

One well-known heuristic is *minimum local fill*, which works on the adjacency graph in the following way: At each stage of elimination, it makes a prediction of the number of new edges that would be created if we were to eliminate this node (that is, the fill-in). The elimination node is then chosen to be one that attains the minimum fill-in.

It is more refined than minimum degree, which essentially counts *all* the elements that are modified by the elimination step, even those for which fill-in does not occur because they were already nonzero. It does not, however, achieve the minimum fill-in for the overall factorization because it looks ahead only to the next elimination step. Minimum local fill is also more expensive to implement than minimum degree, and, in most applications, the extra cost of the ordering is not justified by the reduction in fill-in. In primal-dual codes, however, the nonzero structure of the matrix is the same at every iteration, so the ordering is performed only once, during initialization. Mészáros [90] notes that the cost of ordering is therefore amortized over a large number of numerical factorizations, and he describes an inexact variant of this heuristic that is competitive with minimum degree in his computational tests.

Some recent studies have looked at a different class of orderings known as *nested-dissection* orderings. These methods take a global view of the adjacency graph, rather than the local, short-sighted view of the minimum-degree and minimum-local-fill heuristics. The basic step of nested dissection finds a *node separator*—a subset of nodes that, when removed from the adjacency graph along with their incident edges, breaks the graph into two disconnected pieces. In constructing the node ordering, we ensure that the separator nodes are listed *after* the nodes in the remaining graph. The dissection process is then applied recursively to both pieces of the remaining graph.

Many sophisticated techniques are now available for computing node separators. Preliminary testing has shown that they perform remarkably well on matrices of the form (11.3a) that arise in a number of large linear programming test problems. We refer the reader to Karypis, Gupta, and Kumar [59] and Rothberg and Hendrickson [118] for some details of numerical tests. The first report focuses on parallel computing aspects of nested-dissection orderings (see the Notes and References), whereas the second contains an extensive list of references to literature in the area.

While much testing remains to be done, this recent work on alternatives to minimum-degree orderings may result in a significant advance in the efficiency of interior-point codes on large linear programs.

Small Pivots in the Cholesky Factorization

We noted earlier that the Cholesky algorithm is stable; that is, it gives accurate solutions to well-conditioned problems. The coefficient matrix AD^2A^T in our particular system (11.3a) may not, however, be well conditioned, and we may encounter very small (even zero or negative) diagonal pivot elements during the numerical factorization. In this section, we discuss how the ill conditioning comes about, how the small pivots are handled, and what effect they have on the quality of the step $(\Delta x, \Delta\lambda, \Delta s)$ computed from (11.3).

Since $D^2 = S^{-1}X$ is diagonal, our coefficient matrix is a weighted sum of outer products of the columns of A; that is,

$$AD^2A^T = \sum_{i=1}^{n} \frac{x_i}{s_i} A_{\cdot i} A_{\cdot i}^T. \tag{11.9}$$

As the analysis of preceding chapters has shown and computational experience has verified, the iterates of most primal-dual algorithms tend toward strictly complementary solutions. Hence, if we use the partition $\{1, 2, \ldots, n\} = \mathcal{B} \cup \mathcal{N}$ from (2.11) once again, we obtain

$$x_i^k / s_i^k \to \infty \text{ for } i \in \mathcal{B}, \qquad x_i^k / s_i^k \to 0 \text{ for } i \in \mathcal{N}.$$

For the path-following algorithms of Chapter 5, we have $x_i^k/s_i^k = O(\mu_k^{-1})$ for $i \in \mathcal{B}$ and $x_i^k/s_i^k = O(\mu_k)$ for $i \in \mathcal{N}$ (Lemma 5.13). As we approach the solution, the ith term in the sum (11.9) grows large for each $i \in \mathcal{B}$, so the norm of AD^2A^T approaches ∞ during the later stages of the algorithm. In many instances, this matrix remains well conditioned, since all its eigenvalues grow to ∞ at more or less the same rate. Often, however, the condition number of AD^2A^T deteriorates during later iterations. This behavior may be caused by a number of factors, including the following:

- The basic part of the constraint matrix A—the submatrix $A_{\mathcal{B}}$—may not have full row rank.
- The set $\mathcal{B}$ contains fewer than m indices, so that at least $m - |\mathcal{B}|$ eigenvalues of AD^2A^T approach zero, whereas the others approach ∞.
- The primal solution set Ω_P is unbounded, so some components of x approach ∞. For these components, the ratios x_i/s_i typically grow even faster than the other indices $i \in \mathcal{B}$, creating an imbalance between the different $\mathcal{B}$ terms in the sum (11.9).

Ill conditioning usually reveals itself in the diagonal pivots encountered during the numerical factorization. These can vary widely in magnitude. Some may even be slightly negative (because of roundoff error from earlier factorization steps), causing the Cholesky algorithm to break down when it tries to take the square root.

We need to modify the Cholesky algorithm to ensure that the step $(\Delta x, \Delta\lambda, \Delta s)$ computed in the presence of roundoff error is a good search direction for the interior-point algorithm, even when the matrix AD^2A^T is ill conditioned. Many modification techniques have been proposed and implemented, all motivated by computational experience rather than theoretical considerations.

Perhaps the most obvious modification is to apply the same diagonal pivoting strategies that are used for singular positive semidefinite matrices (see Higham [52, Chapter 7]). At step i of the factorization, to see whether the next pivot element is not too much smaller than the largest pivot encountered so far, we check the following criterion:

$$M_{ii} \geq \texttt{tol} \times \max_{j=1,\ldots,i-1} L_{jj}^2, \tag{11.10}$$

where `tol` is a small positive quantity comparable to unit roundoff error—say, $\texttt{tol} = 10^{-12}$. If the test (11.10) fails, we choose a larger pivot from among the remaining diagonals $M_{i+1,i+1}, \ldots, M_{mm}$ and perform a row/column swap to move the selected pivot to the (i, i) position. If no diagonal element in the remaining matrix satisfies (11.10), we terminate the factorization and set the remaining columns of L to zero. The computed factorization has the form

$$\begin{bmatrix} L_{11} & 0 \\ L_{21} & 0 \end{bmatrix} \begin{bmatrix} L_{11}^T & L_{21}^T \\ 0 & 0 \end{bmatrix} \approx M, \tag{11.11}$$

where L_{11} is square, lower triangular, and nonsingular. An approximate solution to the system $Mz = r$ is obtained by partitioning z and r as

$$z = \begin{bmatrix} z_1 \\ z_2 \end{bmatrix}, \qquad r = \begin{bmatrix} r_1 \\ r_2 \end{bmatrix}$$

(where z_1 and r_1 have the same dimensions as L_{11}) and obtaining z_1 and z_2 by solving the systems

$$L_{11}L_{11}^T z_1 = r_1, \qquad z_2 = 0. \tag{11.12}$$

We investigate this strategy further in the exercises.

Diagonal pivoting is difficult to implement in the context of interior-point codes because it requires row/column permutations to be performed "on the fly," with all the dynamic reorganization of data structures that these operations entail. Strategies that mimic diagonal pivoting without the need for row and column swaps are implemented in a number of codes. The cleanest strategy of this kind simply skips step i of the Cholesky factorization when it encounters a small pivot M_{ii}, and sets the ith column of L to zero. It also sets the ith component of the solution z to zero, just as the elements of z_2 are set to zero in (11.12). By skipping the ith step of the factorization, this strategy is, in effect, approximating all the elements $M_{ii}, M_{i+1,i}, \ldots, M_{mi}$ by zeros. This approximation is a reasonable one, as we see in the exercises.

Another strategy that is essentially equivalent to the skipping strategy is to replace each small pivot M_{ii} with a huge number (say, 10^{128}) and proceed with the elimination step of the factorization. The effect is to replace the diagonal element L_{ii} of the factor L with 10^{64}, whereas the subdiagonals $L_{i+1,i}, \ldots, L_{mi}$ become small multiples of 10^{-64}. The computed solution z will be almost identical to the one obtained from the skipping strategy. This "huge-pivot" approach has the advantage that it can be implemented by making trivial changes to existing sparse Cholesky codes. In the primal-dual codes LIPSOL [162] and PCx [23], this strategy is implemented by changing just a few lines of Fortran source in the sparse Cholesky code of Ng and Peyton [103].

Wright [150] uses the analytical tools of numerical linear algebra to show that the skipping and huge-pivot strategies produce good search directions for primal-dual algorithms even as the duality measure μ becomes very small. This analysis explains, at least in part, the remarkable success of these simple modifications to the Cholesky algorithm in practical primal-dual implementations.

Dense Columns in A

The formulation (11.3) has another potential difficulty: the matrix AD^2A^T may be much denser than A itself, making the Cholesky strategy based on (11.3) an inefficient way to calculate the step. The usual culprits are dense columns in A that contribute dense outer products to the summation (11.9). If, for instance, half the elements in column i are nonzeros, then 25% of the elements in the outer product $A_{\cdot i}A_{\cdot i}^T$ will be nonzero. Therefore the matrix AD^2A^T will be at least 25% dense, even if the other columns of A are sparse.

Fortunately, if there are not too many dense columns, the efficiency of the Cholesky approach can be salvaged. We sketch the modified procedure in the simple case where the density of AD^2A^T is due to the single column $A_{\cdot i}$. In this case, we have

$$AD^2A^T = \sum_{j \neq i} \frac{x_j}{s_j} A_{\cdot j} A_{\cdot j}^T + \frac{x_i}{s_i} A_{\cdot i} A_{\cdot i}^T, \tag{11.13}$$

where the partial sum over $j \neq i$ is a sparse matrix, whereas the outer product $A_{\cdot i} A_{\cdot i}^T$ is dense. The first step is to perform a (sparse) Cholesky factorization on the partial sum to obtain

$$\tilde{L}\tilde{L}^T = P \left(\sum_{j \neq i} \frac{x_j}{s_j} A_{\cdot j} A_{\cdot j}^T \right) P^T. \tag{11.14}$$

The second step makes use of the Sherman–Morrison–Woodbury formula (see Appendix A), which shows how the inverse of a matrix is affected by low-rank perturbations. Applying the formula (A.22) to (11.13), we can find $\Delta\lambda$ that satisfies (11.3a) by solving two systems whose coefficient matrix is the partial sum (11.14). Each of these two solves requires a pair of inexpensive triangular substitutions involving the factor $\tilde{L}$. We leave the details to the exercises.

This combined Cholesky/Sherman–Morrison–Woodbury procedure effectively replaces a single factorization and solve involving the dense matrix AD^2A^T with a factorization and two solves involving the sparse matrix (11.14). Extension of this procedure to $p > 1$ dense columns is straightforward. Again, we split the dense terms from the sum (11.9) and perform Cholesky factorization on just the sparse part. Then, using (A.24), we can compute the solution of (11.3a) at the cost of a single sparse factorization together with $p + 1$ back-substitutions with the sparse factors.

Primal-dual codes typically identify columns of A as "dense" during a presolve stage if their proportion of nonzeros exceeds a user-defined parameter.

The Augmented System Formulation

We turn now to the augmented system form (11.2) of the step equation. The linear system (11.2a) is larger than the normal equations form (11.3a)—dimension $m + n$ instead of m—but it has some advantages in robustness and flexibility over the normal equations form. These advantages include the following:

- Dense columns in A are not as troublesome, since the outer products $A_{\cdot i} A_{\cdot i}^T$ are not computed a priori as in (11.3a). The factorization algorithms for (11.2a) do not need to make special provision for dense columns.

- Ill conditioning occurs in the matrix (11.2a), as in (11.3a), but it is easier to trace the effects on the numerical factorization and, ultimately, on the quality of the solution vector $(\Delta x, \Delta\lambda, \Delta s)$.

- The augmented system form extends more easily to alternative formulations of the linear programming problem (1.1), (1.2). In particular, formulations that contain free variables—components of x for which there are no explicit upper or lower bounds—are accommodated easily by the augmented system form, while the normal equations form requires an awkward reformulation of the problem, as we see below.

Issues of conditioning and stability have received some recent attention in the literature. Wright [148] shows that, despite roundoff error, the steps $(\Delta x, \Delta\lambda, \Delta s)$ obtained by applying standard symmetric indefinite factorization algorithms to (11.2) continue to be good search directions until μ becomes quite small (i.e., μ of the order of unit roundoff). Other studies of stability in similar contexts have been carried out by Forsgren, Gill, and Shinnerl [31] and Ponceleón [106].

The augmented system formulation has some disadvantages, too, that are essentially logistical:

- Algorithms and software for solving sparse symmetric indefinite systems are not as highly developed or as widely available as sparse Cholesky codes.

- It takes more computer time (typically, 50%–100% more) to obtain the step $(\Delta x, \Delta\lambda, \Delta s)$ from the augmented system form than from the Cholesky algorithm applied to (11.3).

The first objection is becoming less significant, thanks to considerable recent work on software in this area. The impetus for this work has come not only from the application to primal-dual methods but also from nonlinear optimization, finite element methods for PDEs, and electrical network modeling, where linear systems similar in form to (11.2a) arise frequently. Vavasis [140] catalogs these applications and proposes an algorithm that meets a certain stability criterion. Recent work on algorithms and software is reported by Duff [27], Ashcraft, Grimes, and Lewis [12]; and Fourer and Mehrotra [33].

The second objection—increased solution time—will probably remain, even if significant advances are made in software. Still, the advantages of the augmented system formulation give it an important role to play in future software for primal-dual methods.

Factoring Symmetric Indefinite Matrices

Most factorizations of a symmetric indefinite matrix T have the form

$$PTP^T = L\hat{D}L^T, \tag{11.15}$$

where

- P is a permutation matrix;
- L is a unit lower triangular matrix (that is, its diagonal elements are all 1);
- $\hat{D}$ is a block diagonal matrix, where each block is either 1×1 or symmetric 2×2.

The best-known algorithms, called *Bunch–Parlett* and *Bunch–Kaufman* after their inventors, eliminate either one or two columns and rows of the matrix at a time, adding to the L and $\hat{D}$ factors and updating the remaining matrix at each elimination step. Each elimination step is performed with a diagonal pivot block selected from the remaining matrix $\bar{T}$—either a single diagonal element of $\bar{T}$, or a 2×2 pivot block whose diagonal entries are also diagonal elements of $\bar{T}$. (More details on pivot selection are given below.) A row/column permutation is then performed to move the pivot block to the upper left of the remaining matrix, to obtain

$$\bar{P}\bar{T}\bar{P}^T = \begin{bmatrix} \bar{E} & \bar{C}^T \\ \bar{C} & \hat{T} \end{bmatrix}, \tag{11.16}$$

where $\bar{E}$ is the 1×1 or 2×2 pivot block. Note that we can factor the matrix (11.16) as

$$\bar{P}\bar{T}\bar{P}^T = \begin{bmatrix} I & 0 \\ \bar{C}\bar{E}^{-1} & I \end{bmatrix} \begin{bmatrix} \bar{E} & 0 \\ 0 & \hat{T} - \bar{C}\bar{E}^{-1}\bar{C}^T \end{bmatrix} \begin{bmatrix} I & 0 \\ \bar{C}\bar{E}^{-1} & I \end{bmatrix}^T.$$

Accordingly, the algorithm includes the column(s) $\bar{C}\bar{E}^{-1}$ in the L factor, includes the pivot block $\bar{E}$ in the $\hat{D}$ factor, and subtracts the term $\bar{C}\bar{E}^{-1}\bar{C}^T$ from the lower right submatrix $\hat{T}$. The entire process is now repeated on the remaining matrix $\hat{T} - \bar{C}\bar{E}^{-1}\bar{C}^T$.

Algorithms that follow this basic framework differ only in the way they choose the pivot block. Unlike the symmetric positive definite case, the ordering process cannot be divorced from the numerical factorization. Stability demands that these two operations be combined, so that pivoting takes place dynamically ("on the fly") during the numerical factorization. We can finesse this issue in our particular application, since the matrices in (11.2a) that are factored at successive interior-point iterations have the same structure and, often, similar numerical values. The code of Fourer and Mehrotra

[33] performs dynamic ordering and factorization during the first interior-point iteration and then reuses the same ordering on later iterations until it encounters numerical difficulties. It then performs another dynamic ordering/factorization step and saves the new ordering for reuse on subsequent iterations.

The key criterion for choosing the pivot block E is that the growth of elements in the remaining matrix $\hat{T} - \bar{C}\bar{E}^{-1}\bar{C}^T$ is not too great. The Bunch–Parlett procedure examines all elements in the remaining matrix and identifies the largest diagonal and largest off-diagonal elements, whose magnitudes are denoted by χ_{diag} and χ_{off}, respectively. The ratio $\chi_{\text{off}}/\chi_{\text{diag}}$ indicates the growth that we can expect in the remaining matrix if the largest diagonal is selected as a 1×1 pivot. If the ratio is not too large, this diagonal element is selected as the pivot. Otherwise, a 2×2 pivot of the form

$$\bar{E} = \begin{bmatrix} \bar{T}_{ii} & \bar{T}_{ij} \\ \bar{T}_{ij} & \bar{T}_{jj} \end{bmatrix}$$

is chosen, where $\bar{T}_{ij}$ is the off-diagonal element that achieves the largest magnitude χ_{off}. It can be shown that the use of $\bar{E}$ will result in only modest growth in the remaining matrix.

The Bunch–Kaufman algorithm achieves similar control over element growth at a lower cost. It examines just two columns of the remaining matrix during its search for a pivot, rather than the entire matrix.

For sparse matrices, a second criterion is also used in selecting a pivot: The update step $\hat{T} \leftarrow \hat{T} - \bar{C}\bar{E}^{-1}\bar{C}^T$ should not create too much fill-in. Fourer and Mehrotra [33] describe a modification of the Bunch–Parlett strategy that combines stability and sparsity considerations. Their strategy determines the number of nonzeros in the update matrices $\bar{C}\bar{E}^{-1}\bar{C}^T$, where $\bar{E}$ is the set of all possible 1×1 and 2×2 pivots in the remaining matrix. It first examines all pivots for which the update matrices have the minimum number of nonzeros and checks to see whether any of them satisfies a stability condition that prevents excessive growth. For a prospective 1×1 pivot, this condition is

$$|\bar{E}^{-1}| \, \|\bar{C}\|_\infty \le \delta^{-1},$$

whereas the 2×2 stability condition is

$$|\bar{E}^{-1}| \begin{bmatrix} \|\bar{C}_{\cdot 1}\|_\infty \\ \|\bar{C}_{\cdot 2}\|_\infty \end{bmatrix} \le \begin{bmatrix} \delta^{-1} \\ \delta^{-1} \end{bmatrix},$$

where δ is a small parameter. If any of the eligible pivots satisfy one of these conditions, they are selected, with a preference given to 1×1 pivots. Otherwise, the procedure examines pivots for which the update matrix $\bar{C}\bar{E}^{-1}\bar{C}^T$ is successively more dense, stopping when it finds a pivot that satisfies one of the stability conditions above.

Starting Points

Theorem 6.1 guarantees global convergence of Algorithm IPF from any starting point (x^0, λ^0, s^0) with $(x^0, s^0) > 0$. Since most of the software is based on algorithms closely related to Algorithm IPF, it appears that we have considerable flexibility in our choice of a starting point. However, the analysis of Chapters 5 and 6 suggests strongly that a good starting point should also satisfy two other conditions. First, the points should be well centered, so that the pairwise products $x_i^0 s_i^0$ are similar for all $i = 1, 2, \ldots, n$. Second, the point should not be *too* infeasible; that is, the ratio $\|(r_b^0, r_c^0)\|/\mu_0$ of infeasibility to duality measure should not be too large. The theory has been borne out by computational experience—good progress can be made only from starting points that satisfy both properties.

A popular heuristic for finding (x^0, λ^0, s^0) starts by calculating $(\tilde{x}, \tilde{\lambda}, \tilde{s})$ as the solution of two least-squares problems:

$$\begin{aligned} \min_x \ \|x\|^2 \quad & \text{subject to } Ax = b, \\ \min_{(\lambda,s)} \ \|s\|^2 \quad & \text{subject to } A^T\lambda + s = c. \end{aligned}$$

That is, $\tilde{x}$ and $\tilde{s}$ are the vectors of least norm for which the residuals r_b and r_c are zero. The starting point is then defined as

$$(x^0, \lambda^0, s^0) = (\tilde{x} + \tilde{\delta}_x e, \tilde{\lambda}, \tilde{s} + \tilde{\delta}_s e),$$

where the scalars $\tilde{\delta}_x$ and $\tilde{\delta}_s$ are calculated from a complicated formula to satisfy the two conditions discussed in the previous paragraph. Details are given in Mehrotra [88, section 7].

Prior information is often available about the solution of a linear program, in the form of a solution of a problem with slightly different data or an estimate of the optimal basis. Primal-dual methods can use this information to construct "hot" starting points, which often yield convergence in fewer iterations than the "cold" starting points discussed above. To construct a hot start from an estimate of the primal-dual solution, we add small positive values to the components of (x, s) that are at or near their lower bound of zero, and possibly make slight adjustments to the larger components of (x, s) as well. These adjustments should be chosen to ensure that the centrality and infeasibility conditions mentioned above are satisfied.

The initial values of μ_0 and $\|(r_b^0, r_c^0)\|$ will be smaller than those obtained from the cold-start heuristic, so that fewer iterations are needed to find a solution. However, primal-dual algorithms do not derive much benefit from hot starting points—possibly a factor of two or three savings in iteration count and run time. The reason for this unfortunate observation is that small perturbations in the problem data (A, b, c) often lead to large changes in the primal-dual solution set Ω due to an interchange of indices between the sets $\mathcal{B}$ and $\mathcal{N}$, causing the central path $\mathcal{C}$ to take a very different course near the solution set (see Figure 7.2). Often, the hot starting point is far from the new path $\mathcal{C}$, and a number of primal-dual iterations are required to move into a closer neighborhood of $\mathcal{C}$ before progress can be made toward a solution.

The simplex method is, on the other hand, much better able to exploit prior information about the solution. Using this information, it can construct an initial basis of m components of x, then perform an initial factorization of the corresponding column submatrix of A (if necessary) and take just a few steps to reach the new solution. Primal-dual codes that incorporate a procedure for recovering an optimal basis/vertex solution such as the one described in Chapter 7 can take advantage of this property of the simplex method. Instead of trying to hot-start the primal-dual algorithm, we can use the optimal basis obtained from the basis recovery procedure to hot-start the simplex algorithm. Hence, to solve a sequence of similar problems, we could use the primal-dual algorithms with optimal basis recovery to solve the first problem in the sequence, then switch to hot-started simplex to solve the remaining problems.

Termination

Unlike the simplex method, primal-dual algorithms never find an exact solution of the linear program. Termination criteria must therefore be used to decide when the current iterate is a close enough to the solution set. The software may also incorporate a finite termination phase that jumps from the approximate solution to an exact solution, or a phase that identifies an optimal basis/vertex solution.

We described both finite termination and optimal basis recovery in Chapter 7, to which we refer the reader for details. These features are not widely implemented in existing primal-dual codes, with the exception of

CPLEX/Barrier and various experimental codes. This neglect is not too surprising, since the code required to implement optimal basis recovery is similar in complexity to a full-blown simplex code.

Most primal-dual codes simply report an approximate solution for which the residuals and duality measure are sufficiently small. Relative measures of smallness are used to lessen the effect of scaling, and a typical test for accepting the point (x, λ, s) as an approximate solution consists of the following three conditions:

$$\frac{\|r_b\|}{1+\|b\|} = \frac{\|Ax-b\|}{1+\|b\|} \leq \epsilon, \tag{11.17a}$$

$$\frac{\|r_c\|}{1+\|c\|} = \frac{\|A^T\lambda + s - c\|}{1+\|c\|} \leq \epsilon, \tag{11.17b}$$

$$\frac{|c^Tx - b^Ty|}{1+|c^Tx|} \leq \epsilon, \tag{11.17c}$$

where ϵ is a small positive number, usually 10^{-8}.

Alternative Formulations for the Linear Program

We mentioned in Chapter 2 that simple transformations can be used to express any linear program in standard form (1.1). Slack variables can be introduced, free variables split into positive and negative parts (2.3), and so on. These transformations produce a coefficient matrix A with a certain amount of structure—rows that contain just one or two nonzeros, for instance—which is subsequently ignored in the formulation of the linear systems (11.2) or (11.3). Most primal-dual implementations avoid forcible transformation to standard form, accepting instead slightly more complicated formulations of the linear program. The KKT system (1.4) and the step equations (11.1) can be adapted to fit the alternative formulations, and, through a process of block elimination, the step equations can also expressed in more compact forms analogous to (11.2) and (11.3). When we use the more complicated formulations, we find that the matrix to be factored at each iteration is generally smaller than if we had transformed to standard form and then used the usual formulations (11.2) or (11.3) above.

To illustrate this point, let us consider an alternative linear programming formulation with upper bounds on the variables:

$$\min\ c^Tx \ \text{ subject to } Ax = b,\ \ 0 \leq x \leq u, \tag{11.18}$$

where u is a nonnegative vector. For simplicity, we assume that *all* the components of x have upper bounds; that is, all components of u are finite. (See the exercises for a variant of (11.18) that does not insist on this condition.) Introducing a slack vector $w \in \mathbb{R}^n$ for the upper bound constraint $x \leq u$ and corresponding Lagrange multiplier vector $\xi \in \mathbb{R}^n$, we can write the KKT conditions for (11.18) as

$$
\begin{aligned}
A^T\lambda + s - \xi &= c, && (11.19a)\\
Ax &= b, && (11.19b)\\
x + w &= u, && (11.19c)\\
x_i s_i &= 0, \quad i = 1, 2, \ldots, n, && (11.19d)\\
w_i \xi_i &= 0, \quad i = 1, 2, \ldots, n, && (11.19e)\\
(x, w, s, \xi) &\geq 0. && (11.19f)
\end{aligned}
$$

In the language of Chapter 8, this problem is an mLCP (see the exercises). For any point (x, λ, s, w, ξ) with $(x, s, w, \xi) > 0$, the general step of a primal-dual method has the following form (cf. (8.8) and (11.1)):

$$
\begin{bmatrix} 0 & A & 0 & 0 & 0 \\ A^T & 0 & I & 0 & -I \\ 0 & I & 0 & I & 0 \\ 0 & S & X & 0 & 0 \\ 0 & 0 & 0 & E & W \end{bmatrix}
\begin{bmatrix} \Delta\lambda \\ \Delta x \\ \Delta s \\ \Delta w \\ \Delta\xi \end{bmatrix}
=
\begin{bmatrix} -r_b \\ -r_c \\ -r_u \\ -r_{xs} \\ -r_{w\xi} \end{bmatrix}, \tag{11.20}
$$

where

$$
r_u = x + w - u, \qquad r_c = A^T\lambda + s - \xi - c,
$$
$$
W = \operatorname{diag}(w_1, w_2, \ldots, w_n), \qquad E = \operatorname{diag}(\xi_1, \xi_2, \ldots, \xi_n).
$$

For the generic primal-dual step (analogous to (1.20)), we would have $r_{w\xi} = WEe - \sigma\mu e$, where μ in this case is defined as

$$
\mu = \frac{x^T s + w^T \xi}{2n}.
$$

Since S, X, W, and E are all diagonal matrices with positive diagonal elements, we can eliminate Δs, Δw, and $\Delta\xi$ from (11.20) to obtain the following analogue of the augmented system form (11.2a):

$$
\begin{bmatrix} 0 & A \\ A^T & -\bar{D}^{-2} \end{bmatrix}
\begin{bmatrix} \Delta\lambda \\ \Delta x \end{bmatrix}
=
\begin{bmatrix} -r_c - W^{-1}r_{w\xi} + X^{-1}r_{xs} + W^{-1}Er_u \\ -r_b \end{bmatrix}, \tag{11.21}
$$

where the diagonal matrix $\bar{D}$ is defined by

$$\bar{D} = (X^{-1}S + W^{-1}E)^{-1/2}.$$

Elimination of Δx finally yields the analogue of the normal equations form (11.3a):

$$A\bar{D}^2 A^T \Delta\lambda = -r_b + A\bar{D}^2 \left(-r_c - W^{-1}r_{w\xi} + X^{-1}r_{xs} + W^{-1}Er_u\right). \quad (11.22)$$

We conclude that, despite the addition of the upper bounds to the formulation (1.1) and the introduction of new variables w and ξ to the primal-dual formulation, the size of the matrix to be factored at each iteration remains the same. Therefore, the time to compute each primal-dual step hardly increases at all. If we took the alternative approach of transforming (11.18) to standard form, the number of rows in the constraint matrix A would grow from m to $m + n$, and the dimensions of the linear systems (11.2a) and (11.3a) would grow accordingly.

Free Variables

In many practical instances of linear programming, it is natural to allow some variables to be *free*, that is, to have no explicit upper or lower bounds. We can modify the standard form to include variables of this type as follows:

$$\min c^T x + d^T z \text{ subject to } Ax + Cz = b, \ x \geq 0, \quad (11.23)$$

where the free variables are denoted by z.

As for the formulation (11.18) with upper bounds, we can write the KKT conditions for (11.23) and formulate the step equations in a number of different ways. The special property of free variables is, however, that they prevent us from deriving a step equation formulation with a positive definite coefficient matrix. That is, there is no analogue to (11.3a) or (11.22) to which we can apply a sparse Cholesky solver. We can obtain a formulation of this type only by reformulating the original problem (11.23) in standard form (1.1) via a splitting of the free variables, but this approach raises some difficulties of its own, as we see below.

The KKT conditions for (11.23) are

$$A^T\lambda + s = c, \quad (11.24a)$$
$$C^T\lambda = d, \quad (11.24b)$$
$$Ax + Cz = b, \quad (11.24c)$$
$$x_i s_i = 0, \quad i = 1, 2, \ldots, n, \quad (11.24d)$$
$$(x, s) \geq 0. \quad (11.24e)$$

Given any primal-dual iterate (x, z, λ, s) with $(x, s) > 0$, the step equation has the following form:

$$\begin{bmatrix} 0 & A & C & 0 \\ A^T & 0 & 0 & I \\ C^T & 0 & 0 & 0 \\ 0 & S & 0 & X \end{bmatrix} \begin{bmatrix} \Delta\lambda \\ \Delta x \\ \Delta z \\ \Delta s \end{bmatrix} = \begin{bmatrix} -r_b \\ -r_c \\ -r_d \\ -r_{xs} \end{bmatrix}, \tag{11.25}$$

where the residual terms r_b, r_c, and r_d are defined in obvious ways. Elimination of Δs produces the augmented system form

$$\begin{bmatrix} 0 & A & C \\ A^T & -D^{-2} & 0 \\ C^T & 0 & 0 \end{bmatrix} \begin{bmatrix} \Delta\lambda \\ \Delta x \\ \Delta z \end{bmatrix} = \begin{bmatrix} -r_b \\ -r_c + X^{-1} r_{xs} \\ -r_d \end{bmatrix}, \tag{11.26}$$

where $D = S^{-1/2} X^{1/2}$ as before. The coefficient matrix in (11.26) is symmetric indefinite, so we can use the LDL^T factorizations described earlier to solve the system, just as we did for (11.2a) and (11.21).

If we now use the diagonal matrix D^{-2} to eliminate the columns of A from (11.26), we obtain

$$\begin{bmatrix} AD^2A^T & C \\ C^T & 0 \end{bmatrix} \begin{bmatrix} \Delta\lambda \\ \Delta z \end{bmatrix} = \begin{bmatrix} -r_b + AD^2(-r_c + X^{-1} r_{xs}) \\ -r_d \end{bmatrix}. \tag{11.27}$$

Since there is no way to eliminate the matrix C in general, this system is the closest we can come to the normal equations form (11.3a). Unfortunately, it is still indefinite (the number of negative eigenvalues equals the dimension of z), so we cannot apply the Cholesky algorithm to it. We could still use the LDL^T factorizations described earlier, but for reasons of stability and sparsity preservation, it is better to apply these techniques directly to the augmented system (11.26).

To reformulate (11.23) in standard form, we simply split the free vector z into positive and negative parts z^+ and z^- and make the substitution

$$z = z^+ - z^-, \qquad (z^+, z^-) \geq 0.$$

The formulation (11.23) becomes

$$\begin{aligned} &\min\ c^T x + d^T z^+ - d^T z^- \\ &\text{subject to } Ax + Cz^+ - Cz^- = b, \ \ (x, z^+, z^-) \geq 0. \end{aligned} \tag{11.28}$$

Since (11.28) is in standard form, we can formulate the step equations in either augmented system form (11.2) or normal equations form (11.3) by making the substitutions

$$A \leftarrow \left[\begin{array}{ccc} A & C & -C \end{array} \right], \quad x \leftarrow (x, z^+, z^-), \quad c \leftarrow (c, d, -d).$$

Primal-dual codes based on Cholesky factorizations usually use this reformulation of the problem, but it has a disadvantage: If the original problem (11.23) has a solution, the solution set for the reformulated problem is unbounded. This fact is easy to verify: If (x, z^+, z^-) is a solution of (11.28), then so is $(x, z^+ + f, z^- + f)$ for any nonnegative vector f. When faced with an unbounded solution set, the iterates of most primal-dual algorithms tend to blow up. In the case of (11.28), both z_i^+ and z_i^- usually approach $+\infty$ for each i, whereas their difference $z_i^+ - z_i^-$—the free variable z_i—converges to a constant value. As we mentioned in the discussion on small pivots in the Cholesky factorization, this behavior causes instability in the linear algebra, since the corresponding terms dominate the matrix AD^2A^T (11.9), swamping the contribution from the non-free-variable terms.

One fairly effective remedy for this tendency to blow up is to adjust the split free variables after each iteration, reducing them to a moderate size while keeping their differences $z_i^+ - z_i^-$ constant. In other words, we set

$$z_i^+ \leftarrow z_i^+ - \eta, \qquad z_i^- \leftarrow z_i^- - \eta,$$

choosing $\eta \in [0, \min(z_i^+, z_i^-))$ so that the smaller of the two values z_i^+ and z_i^- is shifted to a prescribed constant.

The tendency to blow up is also ameliorated by the use of a superlinear algorithm. Such algorithms use search directions close to the affine-scaling direction during their later iterations, and, as we see from Theorem 7.11, they produce iteration sequences that converge. This feature is observed in practice: Superlinear codes are definitely more robust when applied to problems with unbounded solution sets than nonsuperlinear codes.

Presolving

The formulations of many practical linear programs contain infelicities—variables defined but never used, rows of A containing only zeros, and so on. Sometimes, the awkwardness in the formulation is quite subtle, as when a combination of constraints on a particular subset of the variables forces all these variables to take on certain fixed values. Most linear programming codes include a *presolver* (also known as a *preprocessor*), whose purpose is to detect and eliminate many of these features from the data prior to activating the linear programming algorithm. By reducing the size of the problem and eliminating features that could lead to numerical difficulties, presolvers benefit simplex and interior-point codes alike.

A few of the checks performed by a presolver are as follows:

Zero rows and columns in A: When the column $A_{\cdot i}$ is zero, the value of x_i has no effect on the product Ax. We can deduce the value of x_i by considering its bounds and the ith element c_i of the cost vector. If $c_i < 0$, then the product $c_i x_i$ is minimized by setting x_i to its upper bound; if there *is* no upper bound, the primal objective must be unbounded below. Similarly, when $c_i > 0$, x_i is forced to its lower bound. When $c_i = 0$, x_i is irrelevant to the problem.

Duplicate rows and columns in A: When two columns of A are simply scalar multiples of one another, we can replace the two components of x with a single primal variable and reduce n by 1. Similarly, when A contains duplicated rows, we can combine two components of λ and reduce m by 1.

Row singleton: When the ith row of A contains a single nonzero element A_{ij}, we can immediately set $x_j = b_i/A_{ij}$ and eliminate row i and column j from the problem, reducing both m and n by 1.

Forced row: Sometimes the combination of a single constraint and bounds forces a collection of variables to take on certain fixed values. For instance, if one of the constraints is $10x_3 - 4x_{10} + x_{12} = -4$, and the bounds on the variables are

$$x_3 \in [0, +\infty), \qquad x_{10} \in [0, 1], \qquad x_{12} \in [0, +\infty),$$

the values of these three variables must be $x_3 = 0$, $x_{10} = 1$, and $x_{12} = 0$. We can eliminate all three variables and the corresponding row of A from the linear program.

Other presolve operations can include eliminating or tightening some bounds, or combining some of the rows of A to improve sparsity in the matrix AD^2A^T from (11.3a).

Presolving is usually implemented by making several sweeps through the rows and columns of A. When we detect an opportunity to reduce the problem, we flag the appropriate rows and columns as "deleted" and push the information onto a stack. One sweep through A can uncover new opportunities for reduction, so we perform repeated passes through the data until no more reductions are found. Once the reduced problem is solved, a postprocessing phase is activated to undo the work of the presolver, popping the stack to express the solution in terms of the original model.

Presolving is generally much less expensive than a single primal-dual iteration, so it is almost always worth doing. More information about presolving can be found in the papers by Andersen and Andersen [5] and Andersen et al. [6] and in their citation lists.

Primal-Dual Codes

The recent paper of Andersen et al. [6] tests a number of primal-dual codes alongside the state-of-art-simplex code, CPLEX/Simplex, on six large linear programs. The primal-dual codes, which include CPLEX/Barrier, LIPSOL, LOQO, HOPDM, and BPMPD, are roughly similar in performance, and they are all faster than CPLEX/Simplex on three of the six problems. This test is hardly comprehensive, but it does reflect the widespread consensus that interior-point methods have a vital role to play in solving large-scale problems. More specific conclusions regarding interior-point versus simplex are not appropriate at this stage. Codes for both classes of method have taken giant strides in recent years, and it is possible that further significant advances lie ahead. The ultimate conclusion is likely to be that the context and the structure of each problem will determine whether an interior-point method, a simplex method, or a hybrid of both is the most appropriate solution technique.

See Appendix B for information about current software packages and relevant sites on the World Wide Web.

Notes and References

The book of George and Liu [38] is a standard reference on sparse Cholesky algorithms and ordering heuristics. A paper by the same authors [39] gives an overview of the minimum-degree ordering and its variants.

The strategies for dealing with small diagonal elements in the Cholesky factorization can also be viewed in terms of the QR factorization of the matrix DA^T. The upper triangular matrix R that satisfies $QR = DA^T$ (Q orthogonal) is simply the transpose of the Cholesky factor L for which $LL^T = AD^2A^T$, provided that we restrict both R and L to have positive diagonals. Symmetric row/column permutation of the matrix AD^2A^T is equivalent to column permutation of DA^T, so the diagonal pivoting strategy described above is equivalent to QR factorization with column pivoting. In estimating the rank of AD^2A^T to be the number of nonsmall pivots, we are assuming that QR factorization with column pivoting effectively reveals the numerical rank of A—an assumption that is usually justified but not foolproof (see Lawson and Hanson [72, p. 31]).

The procedure that we describe for handling dense columns in A can also be viewed in another way. This procedure effectively reformulates the step equations as a hybrid of (11.2) and (11.3), in which we eliminate only those components of Δx from (11.2a) that correspond to sparse columns of A. In the case of a single dense column $A_{\cdot i}$ (as in (11.13)), the result is a linear system whose coefficient matrix is

$$\begin{bmatrix} \sum_{j\neq i} \frac{x_j}{s_j} A_{\cdot j} A_{\cdot j}^T & A_{\cdot i} \\ A_{\cdot i}^T & -(s_i/x_i) \end{bmatrix}. \tag{11.29}$$

We can then solve this system by finding the Cholesky factor of the $(1,1)$ block (as in (11.14)) and use this factor to construct the Schur complement at the cost of a single back-substitution with right-hand side $A_{\cdot i}$. The factorization of the entire matrix (11.29) follows trivially. This procedure is equivalent to (and costs the same as) the Sherman–Morrison–Woodbury–based procedure discussed above.

We have assumed in both cases that the partial sum in (11.14) is safely positive definite, so that its Cholesky factorization can be computed without difficulty. In fact, the partial sum may have much poorer conditioning than the full system AD^2A^T, in which case the techniques just described may produce inaccurate steps. Andersen [8] stabilizes the process by adding nontrivial values to any small pivots encountered during the factorization of the partial sum (11.14). Since these additions represent a low-rank perturbation of the matrix, the Sherman–Morrison–Woodbury formula (A.24) must be used to adjust the resulting factors.

The primal-dual code in OSL, called EKKBSLV, uses a somewhat different technique to handle the dense columns. It applies the iterative method LSQR to the system (11.3a), using the Cholesky factorization of the sparse part of ADA^T to construct a preconditioner. LSQR is a well-known method due to Paige and Saunders [105] for sparse linear least-squares problems which, in exact arithmetic, is identical to conjugate gradient on the normal equations (11.3a). After preconditioning, the coefficient matrix is a low-rank perturbation of the identity matrix, so standard theory for the conjugate gradient algorithm implies convergence in just a few iterations [42, Theorem 10.2.4]. LIPSOL [162] includes this approach as an option.

The role of iterative linear equations solvers in interior-point methods has been a source of curiosity from the earliest days. Many computational experiments have been tried since then, but little success has been reported on general linear programs. (As a general rule, negative results are rarely reported.) The obvious approach would be to apply a preconditioned conjugate gradient method to a scaled form of (11.3a), but experiments have shown that such methods are much slower on typical problems than are direct Cholesky factorizations. Another reasonable approach has been tried by Freund and Jarre [35], who apply the quasi-minimum residual (QMR) algorithm to the augmented system, using an incomplete Bunch–Parlett preconditioner. Their computational results are inconclusive.

Iterative methods have, however, met with success on a special, but very important, class of problems: network linear programs. Some recent papers have described conjugate gradient methods for the system (11.3a), with preconditioners based on the underlying network structure of the problem. Because these iterative methods produce only inexact solutions of the linear equations, some of the interior-point convergence theory needs to be reassessed. We mention in particular the reports by Mehrotra and Wang [89] and Portugal et al. [107].

Many practical linear programs are extremely large, and turnaround in real time is necessary or at least highly desirable. These factors make linear programming a legitimate application for parallel implementation. Interior-point methods are better suited to implementation on advanced multiprocessors than is the simplex method because the amount of computation per iteration—the unit of work to be divided between the processors—is much greater for interior-point methods.

A parallel primal-dual implementation, CPLEX Parallel Barrier, is available already as a commercial product (Lustig and Rothberg [80]). This code executes on a Silicon Graphics Power Challenge, a shared-memory multiprocessor. The code is an implementation of the Mehrotra predictor-corrector algorithm, using the normal equations form of the step equations (11.3), in which all the major arithmetic operations have been carefully parallelized. The Cholesky factorization is the most difficult operation to parallelize. Amalgamation of similar supernodes is performed to coarsen the granularity of the computation. The other operations—formation of the matrix AD^2A^T, triangular substitution with L and L^T, computation of the maximum step lengths $\alpha_{\max}^{\text{pri}}$ and $\alpha_{\max}^{\text{dual}}$ from (10.9)—tend to take less computer time, but they too must be parallelized to avoid creating bottlenecks. Overall performance of the parallel implementation is impressive—sustained performance of over 2 gigaflops on one large problem is reported.

Karypis, Gupta, and Kumar [59] describe a parallel implementation with an emphasis on scalability of the sparse Cholesky algorithm and report computational results for this factorization on up to 256 multiprocessor nodes. Parallelism in their Cholesky implementation is achieved by eliminating many columns concurrently. It is easy to see from our description of the Cholesky factorization that column i is eligible for elimination as soon as the updates to this column due to earlier elimination steps have been accounted for. (In general, there is no need to wait until column $i - 1$ has been eliminated.) At any given time, more than one column is eligible for elimination according to this criterion; all such columns can be eliminated concurrently on different processors. Effective node orderings for parallel sparse Cholesky aim to keep the number of columns eligible for elimination at a consistently high level throughout the factorization. In [59], the authors experiment with minimum-degree and nested-dissection orderings. They show that the nested-dissection orderings tend to create more fill-in but become competitive on large numbers of processors because they are more amenable to parallelism. However, the differences between the two approaches do not appear to be dramatic.

Exercises

1. Given that X and S are diagonal with positive diagonal elements, show that the matrix M in (11.4) is symmetric and positive definite if and only if A has full row rank. Does this result still hold if we replace $S^{-1}X$ by a diagonal matrix in which exactly m of the diagonal elements are positive and the remainder are zero? Explain.

2. Show that the matrix in (11.2a) is indefinite.

3. If M is a symmetric positive definite matrix, show that none of its diagonal elements can be nonpositive. Show also that after each step of Cholesky factorization, the remaining matrix is also symmetric positive definite.

4. Formula (11.7) depicts the fill-in at the first step of the Cholesky factorization for a simple example. Would additional fill-in be observed at the second or third steps?

5. Suppose that we have a square matrix L of the form

$$L = \begin{bmatrix} L_{11} & 0 \\ L_{21} & 0 \end{bmatrix},$$

where L_{11} is square and nonsingular, and that we seek a solution to the consistent system of equations $LL^T z = d$, written in partitioned form as

$$\begin{bmatrix} L_{11} & 0 \\ L_{21} & 0 \end{bmatrix} \begin{bmatrix} L_{11}^T & L_{21}^T \\ 0 & 0 \end{bmatrix} \begin{bmatrix} z_1 \\ z_2 \end{bmatrix} = \begin{bmatrix} d_1 \\ d_2 \end{bmatrix}.$$

   (i) Show that consistency of the system implies that $d \in \text{Range}(L)$ and that $d_2 = L_{21}L_{11}^{-1}d_1$.

   (ii) Show that any solution of $LL^T z = d$ has the form

$$z_1 = L_{11}^{-T}(L_{11}^{-1}d_1 - L_{21}^T z_2), \qquad z_2 \quad \text{arbitrary.}$$

6. Let M be an $m \times m$ symmetric positive semidefinite matrix such that $\max_{j=1,2,\ldots,m} M_{jj} = 1$ and $M_{11} = \delta$, where δ is small and positive. Show that the elements $M_{21}, M_{31}, \ldots, M_{m1}$ are all bounded in magnitude by $\delta^{1/2}$. Hint: For any vector $x \in \mathbb{R}^m$ of the form $x = (1, 0, \ldots, 0, t, 0, \ldots, 0)^T$, where the t occurs in position j, we have

$$x^T Mx = M_{11}^2 + 2tM_{j1} + t^2 M_{jj}^2 = \begin{bmatrix} 1 \\ t \end{bmatrix}^T \begin{bmatrix} M_{11} & M_{j1} \\ M_{j1} & M_{jj} \end{bmatrix} \begin{bmatrix} 1 \\ t \end{bmatrix} \geq 0.$$

7. Describe how the formula (A.22) can be used in conjunction with the partial Cholesky factorization (11.13), (11.14) to solve the system (11.3a).

8. Verify the derivation of the linear systems (11.21) and (11.22) from the unreduced form (11.20).

9. (i) Verify the correctness of the KKT conditions (11.19) and (11.24) by applying Theorem A.1 to (11.18) and (11.23), respectively.

   (ii) By making the appropriate identifications with the data in (8.4), verify that (11.19) and (11.24) are mixed monotone LCPs.

   (iii) Write the KKT conditions for the following linear program, in which not all the primal variables (x, z) have upper bounds:

$$\min\ c^T x + d^T z \ \text{ subject to } Ax + Cz = b,\ 0 \leq x \leq u,\ 0 \leq z.$$

10. Write the dual problem for (11.28), and verify that it has no strictly feasible points. (This linear program illustrates Theorem 2.3, which relates strict feasibility of the dual to boundedness of the primal solution set.)

Appendix A

Basic Concepts and Results

In this appendix, we describe some basic concepts and results from optimization, numerical linear algebra, computer science, and complexity theory.

In the case of theorems, we usually state the results and give references for the proofs. We also include the proofs of some results from this book—Lemmas 4.1 and 7.2 and Theorem 9.1—that were omitted from the main text because of their technical nature.

Order Notation

Order notation saves a lot of messy detail when analyzing the asymptotic properties of sequences. We use two varieties of order notation in this book: $O(\cdot)$ and $o(\cdot)$. Given two nonnegative infinite sequences of scalars $\{\eta_k\}$ and $\{\nu_k\}$, we say that

$$\eta_k = O(\nu_k)$$

if there is a positive constant C such that

$$\eta_k \leq C\nu_k$$

for all k sufficiently large. That is, the elements of the first sequence $\{\eta_k\}$ do not become too much larger than the elements of the second sequence $\{\nu_k\}$ as k increases. We write

$$\eta_k = o(\nu_k)$$

if there is another nonnegative sequence $\{\beta_k\}$ such that

$$\eta_k \leq \beta_k \nu_k, \qquad \beta_k \to 0.$$

Convex Sets and Functions

The set $U \subset \mathbb{R}^N$ is *convex* if

$$u, v \in U \quad \Rightarrow \quad tu + (1-t)v \in U \quad \text{for all } t \in [0, 1].$$

Given a function $f : U \to \mathbb{R}$ (where U is convex), we say that f is *convex* if

$$u, v \in U \quad \Rightarrow \quad f(\alpha u + (1-\alpha)v) \le \alpha f(u) + (1-\alpha) f(v) \quad \text{for all } \alpha \in [0, 1]. \tag{A.1}$$

We say that f is *strictly convex* if the inequality in (A.1) is strict when $u \neq v$ and α lies in the open interval $(0, 1)$.

Linear functions (which have the form $f(u) = c^T u + \beta$) are convex.

If U is an open convex set and f is twice continuously differentiable on U with a Hessian that is positive semidefinite for all $u \in U$, then the function f is convex. If the Hessian of f is positive definite for all $u \in U$, then f is strictly convex.

KKT Conditions

Consider the linearly constrained optimization problem

$$\min_u f(u) \quad \text{subject to } Cu = d,\ Gu \ge h,\ u \in U, \tag{A.2}$$

where f is a smooth function, $U \subset \mathbb{R}^N$ is an open set, C and G are matrices, and d and h are vectors. We say that $u^* \in \mathbb{R}^N$ is a *local solution* of (A.2) if

- it is feasible for (A.2)—that is, $u^* \in U$, $Cu^* = d$, and $Gu^* \ge h$;
- there is a scalar $\rho > 0$ such that $f(u) \ge f(u^*)$ for all feasible u with $\|u - u^*\| < \rho$.

The point u^* is a *strict local solution* if it is a local solution and, in addition, $f(u) > f(u^*)$ for all $u \neq u^*$ with u feasible and $\|u - u^*\| < \rho$.

These definitions can be extended to *global solution* and *strict global solution* essentially by taking $\rho = \infty$ in the local definition.

The first-order necessary conditions, also known as the *Karush–Kuhn–Tucker*, or *KKT*, conditions, are defined in the following theorem.

Theorem A.1 *Suppose that u^* is a local solution of* (A.2) *and that f is differentiable in a neighborhood of u^*. Then there are vectors y and z such* *that the following conditions hold:*

$$\begin{aligned}
\nabla f(u^*) - C^T y - G^T z &= 0, & \text{(A.3a)}\\
Cu^* &= d, & \text{(A.3b)}\\
Gu^* &\geq h, & \text{(A.3c)}\\
z &\geq 0, & \text{(A.3d)}\\
z^T(Gu^* - h) &= 0. & \text{(A.3e)}
\end{aligned}$$

The vectors y and z are *Lagrange multipliers.* For a proof of Theorem A.1, see Fletcher [30, Chapter 9] or Mangasarian [81, Chapter 7].

Recalling the definition of strict complementarity of primal-dual solutions for linear programs (see Theorem 2.4 and the discussion that follows this result), we extend the concept of strict complementarity to the more general problem (A.2) in the following definition.

Definition 1 *A solution u^* of the problem* (A.2) *and the corresponding Lagrange multipliers (y, z) that satisfy* (A.3) *are* strictly complementary *if $z + (Gu^* - h) > 0$.*

This definition, in conjunction with the conditions (A.3c), (A.3d), and (A.3e), implies that for any index i we have either

$$z_i = 0, \qquad (Gu^* - h)_i > 0,$$

or

$$z_i > 0, \qquad (Gu^* - h)_i = 0.$$

When we replace the linear constraints in (A.2) with nonlinear constraints, the KKT conditions do not change much. The matrices C and G in (A.3a) are replaced by the Jacobians of the nonlinear constraint functions, and the conditions (A.3b) and (A.3c) are replaced with the nonlinear feasibility conditions. In the nonlinear case, however, the optimality results require another condition known as *constraint qualification.* This condition ensures that the geometry of the feasible region is adequately captured by a linearization of the constraints around u^*. Many different constraint qualifications have been proposed; see Fletcher [30, Chapter 9] and Nocedal and Wright [104] for discussions of this complex issue.

Convexity and Global Solutions

If U is an open convex set (possibly the entire space $\mathbb{R}^N$), then the feasible region for (A.2) is a convex set. If the objective function f is also convex on the feasible set, then we call (A.2) a *convex programming* problem. For convex programming problems, local and global solutions are related in the following way.

Theorem A.2

(i) *If* (A.2) *is a convex programming problem, then the KKT conditions are* sufficient *for u^* to be a global solution. That is, if there are vectors y and z such that* (A.3) *are satisfied, then u^* is a global solution of* (A.2).

(ii) *If, in addition, the function f is* strictly *convex on its feasible region, then any local solution is a uniquely defined global solution.*

Proof. We prove (i) and leave (ii) as an exercise.

Given u^* together with the vectors y and z that satisfy (A.3), we aim to show that

$$f(u^* + v) \geq f(u^*) \tag{A.4}$$

for all v such that $u^* + v$ is feasible.

Because f is convex, we have for any vector v with $u^* + v \in U$ that

$$f(u^* + \alpha v) \leq (1 - \alpha) f(u^*) + \alpha f(u^* + v)$$

and hence

$$\frac{1}{\alpha}\left(f(u^* + \alpha v) - f(u^*)\right) \leq f(u^* + v) - f(u^*).$$

By taking the limit as $\alpha \downarrow 0$ and using differentiability of f, we obtain

$$f(u^* + v) \geq f(u^*) + v^T \nabla f(u^*). \tag{A.5}$$

Since the feasible set is convex, we can express any feasible point as $u^* + v$ for some choice of v. Since $Cu^* = d$ and $C(u^* + v) = d$, the vector v must satisfy $Cv = 0$. Let us also look at the *active* components of the inequality constraint $Gu^* \geq h$; that is, the indices i for which $G_{i\cdot}u^* = h_i$. Because $u^* + v$ is feasible, we must have $G_{i\cdot}v \geq 0$. For the *inactive* components—the rows i for which $G_{i\cdot}u^* > h_i$—we have from conditions (A.3d) and (A.3e) that $z_i = 0$.

We now insert the vector v into the formula (A.5). From (A.3a), we obtain

$$\begin{aligned} v^T \nabla f(u^*) &= v^T (C^T y + G^T z) \\ &= y^T C v + \sum_{i \text{ active}} z_i G_{i \cdot} v + \sum_{i \text{ inactive}} z_i G_{i \cdot} v. \end{aligned}$$

Since $Cv = 0$ and $z_i = 0$ for i inactive, the first and third terms on the right-hand side are zero. The middle term is nonnegative, since z_i and $G_i v$ are both nonnegative when i is an active index. Hence $v^T \nabla f(u^*) \geq 0$, so we have from (A.5) that (A.4) holds.

Hence u^* is a global solution, as claimed. □

Self-Duality and a Proof of Theorem 9.1

Let us return to the self-dual linear program that we discussed in Chapter 9, namely,

$$\begin{array}{lrcll} & \min_{u,w} \; f^T u + g^T w & & & \\ \text{subject to} & M_{11} u + M_{12} w & \geq & -f, & \quad u \geq 0, \; w \text{ free}, \\ & -M_{12}^T u + M_{22} w & = & -g, & \end{array} \qquad \text{(A.6)}$$

where the matrices M_{11} and M_{22} are square and skew-symmetric. Here, we fill in some of the details that were omitted from the discussion in Chapter 9 and prove Theorem 9.1.

By applying the KKT conditions (Theorem A.1) and using skew-symmetry, we find that a solution (u, v) of (A.6) must satisfy

$$\begin{aligned} M_{11}\bar{u} + M_{12}\bar{w} &\geq -f, && \text{(A.7a)} \\ -M_{12}^T \bar{u} + M_{22}\bar{w} &= -g, && \text{(A.7b)} \\ M_{11} u + M_{12} w &\geq -f, && \text{(A.7c)} \\ -M_{12}^T u + M_{22} w &= -g, && \text{(A.7d)} \\ u^T \left[M_{11}\bar{u} + M_{12}\bar{w} + f \right] &= 0, && \text{(A.7e)} \\ \bar{u}^T \left[M_{11} u + M_{12} w + f \right] &= 0, && \text{(A.7f)} \\ u \geq 0, \; w \text{ free}, \; \bar{u} \geq 0, \; \bar{w} \text{ free}, & && \text{(A.7g)} \end{aligned}$$

where $(\bar{u}, \bar{w})$ is a vector of Lagrange multipliers. Note the redundancy in this system. The equation pairs (A.7a)/(A.7c), (A.7b)/(A.7d), and (A.7e)/(A.7f) are duplicates of each other, except for the change of variables $(u, w) \leftrightarrow (\bar{u}, \bar{w})$.

As we noted in Chapter 9, the dual problem is exactly the same as (A.6). Therefore, any solution of the primal is a solution of the dual, so we can set $(\bar{u}, \bar{w}) = (u, w)$ in (A.7) and eliminate the redundancies to obtain

$$\begin{aligned} M_{11}u + M_{12}w &\geq -f, \\ -M_{12}^T u + M_{22}w &= -g, \\ u^T\left[M_{11}u + M_{12}w + f\right] &= 0, \\ u \geq 0,\ w \text{ free.} & \end{aligned}$$

By introducing the variable v, we obtain the mixed monotone linear complementarity (mLCP) formulation that was defined in (9.2):

$$\begin{bmatrix} v \\ 0 \end{bmatrix} = \begin{bmatrix} M_{11} & M_{12} \\ -M_{12}^T & M_{22} \end{bmatrix} \begin{bmatrix} u \\ w \end{bmatrix} + \begin{bmatrix} f \\ g \end{bmatrix}, \qquad (u, v) \geq 0, \quad u^T v = 0. \quad \text{(A.8)}$$

By construction of the mLCP, the vector $(u, v, w) = (u, M_{11}u + M_{12}w + f, w)$ is a solution of (A.8) whenever (u, w) is a solution of (A.6). The converse claim—if (u, v, w) solves (A.8), then (u, w) solves (A.6)—is easily verified by setting the Lagrange multipliers in the KKT conditions (A.7) to $(\bar{u}, \bar{w}) = (u, w)$.

The equivalence between (A.6) and (A.8) is the key to efficient solution of self-dual problems. If we were to apply a primal-dual method naively to (A.6), the unknowns would be u, w, $\bar{u}$, $\bar{w}$, together with the slack variables v and $\bar{v}$. But because of the equivalence with (A.8), we see that it is enough to solve a constrained nonlinear system with half as many variables, namely,

$$F_{\text{mLCP}}(u, v, w) = \begin{bmatrix} f + M_{11}u + M_{12}w - v \\ g - M_{12}^T u + M_{22}w \\ UVe \end{bmatrix} = 0, \qquad (u, v) \geq 0, \qquad \text{(A.9)}$$

where $U = \text{diag}(u_1, u_2, \ldots)$ and $V = \text{diag}(v_1, v_2, \ldots)$.

In Chapter 8, we noted that LCPs, unlike linear programs, do not always have strictly complementary solutions. We now show that the particular mLCP (A.8) *does* have strictly complementary solutions, as claimed in Theorem 9.1.

Proof of Theorem 9.1. The definition of strict complementarity—Definition 1—says that the solution $(u, w, \bar{u}, \bar{w})$ of (A.7) is strictly complementary if

$$u + (M_{11}\bar{u} + M_{12}\bar{w} + f) > 0, \qquad \bar{u} + (M_{11}u + M_{12}w + f) > 0. \quad \text{(A.10)}$$

Since (A.6) and its (self-)dual are feasible, Theorem 2.4 states that there is a strictly complementary primal-dual solution $(u, w, \bar{u}, \bar{w})$ that satisfies (A.7) and (A.10). Therefore, the rearranged vector $(\bar{u}, \bar{w}, u, w)$ is also a strictly complementary primal-dual solution, while (u, w, u, w) is also a solution. We now show that (u, w, u, w) is strictly complementary.

Theorem 2.4 and the discussion surrounding this result (adapted to the nonstandard form of the self-dual problem) imply that for *all* strictly complementary solutions $(u, w, \bar{u}, \bar{w})$, the zero and nonzero elements of the first component u appear in the same locations. Hence, since $(u, w, \bar{u}, \bar{w})$ and $(\bar{u}, \bar{w}, u, w)$ are both strictly complementary, we can identify a subset of indices $\mathcal{B}$ and its complement $\mathcal{B}^c$ such that

$$\begin{aligned}
i \in \mathcal{B} \quad &\Rightarrow \quad u_i > 0, \qquad \bar{u}_i > 0, \\
&\qquad (M_{11}u + M_{12}w + f)_i = (M_{11}\bar{u} + M_{12}\bar{w} + f)_i = 0; \\
i \in \mathcal{B}^c \quad &\Rightarrow \quad u_i = \bar{u}_i = 0, \\
&\qquad (M_{11}u + M_{12}w + f)_i > 0, \qquad (M_{11}\bar{u} + M_{12}\bar{w} + f)_i > 0.
\end{aligned}$$

It follows immediately from these relations that

$$u_i + (M_{11}u + M_{12}w + f)_i > 0 \qquad \text{for all } i, \tag{A.11}$$

so the solution (u, w, u, w) is also strictly complementary.

The solution of (A.8) that corresponds to (u, w, u, w) is $(u, v, w) = (u, M_{11}u + M_{12}w + f, w)$. This solution is a strictly complementary solution for the problem (A.8), since $u + v > 0$, so our proof is complete. □

The Natural log Function and a Proof of Lemma 4.1

In Lemma 4.1, we proposed some simple polynomial approximations to the log function that were essential to the analysis of Chapter 4. We left the proof of Lemma 4.1(i) to the exercises of this chapter. The claim of part (ii) is that, for any $t \in \mathbb{R}^n$ with $\|t\|_\infty \le \tau < 1$, we have

$$-\sum_{i=1}^{n} \log(1 + t_i) \le -e^T t + \frac{\|t\|^2}{2(1-\tau)}. \tag{A.12}$$

To prove this result, we use standard tools: Taylor's theorem and elementary integration. For technical purposes, we introduce the function $f(t)$ defined as

$$f(t) = -\sum_{i=1}^{n} \log(1 + t_i).$$

This function is well defined and smooth for $\|t\|_\infty \le \tau < 1$, and we have

$$f'(t) = \left[-\frac{1}{1+t_i} \right]_n^{i=1}, \qquad f''(t) = \operatorname{diag}\left[\frac{1}{(1+t_i)^2} \right]_n^{i=1}.$$

By Taylor's theorem, we have

$$f(t) = f(0) + t^T f'(0) + \frac{1}{2}\int_0^1 t^T f''(\beta t) t \, d\beta.$$

Now,

$$t^T f''(\beta t) t = \sum_{i=1}^n \frac{t_i^2}{(1+\beta t_i)^2} \le \sum_{i=1}^n \frac{t_i^2}{(1-\beta\tau)^2} = \frac{\|t\|^2}{(1-\beta\tau)^2}.$$

Elementary integration then yields

$$\int_0^1 t^T f''(\beta t) t \, d\beta \le \|t\|^2 \int_0^1 \frac{d\beta}{(1-\beta\tau)^2} = \|t\|^2 \frac{1}{\tau} \left[\frac{1}{1-\beta\tau} \right]_{\beta=0}^{\beta=1} = \frac{\|t\|^2}{1-\tau},$$

and the result (A.12) follows.

Singular Value Decomposition, Matrix Rank, and QR Factorization

Given a real $p \times q$ matrix B with $p \ge q$, there are

- a $p \times p$ orthogonal matrix U,
- a $q \times q$ orthogonal matrix V,
- a $q \times q$ diagonal matrix Σ, with nonnegative diagonal elements arranged in decreasing order

such that

$$B = U \begin{bmatrix} \Sigma \\ 0 \end{bmatrix} V^T. \tag{A.13}$$

The diagonals of Σ are the *singular values*, and (A.13) is the *singular value decomposition* (SVD).

The SVD reveals a great deal about the matrix B. For instance, the *rank* of B is simply the number of nonzero diagonal elements in Σ; that is,

$$\Sigma_{11} > \Sigma_{22} > \cdots > \Sigma_{rr} > 0, \qquad \Sigma_{r+1,r+1} = \cdots = \Sigma_{qq} = 0,$$

where $r = \text{rank}(B)$. The matrix B is said to be *rank deficient* if $\text{rank}(B) < q$.

If B is square and nonsingular, there is no zero block in the middle factor of (A.13), and the inverse of B is

$$B^{-1} = V\Sigma^{-1}U^T.$$

The definitions of matrix norm and condition number can be restated in terms of the quantities in (A.13). Since

$$\|B\| = \|B\|_2 \stackrel{\text{def}}{=} \max_{\|x\|_2=1} \|Bx\|_2, \tag{A.14}$$

it is easy to check that

$$\|B\| = \Sigma_{11}.$$

The *condition number* of B is defined as the ratio of the largest to smallest diagonal elements in Σ; that is,

$$\text{cond}(B) = \Sigma_{11}/\Sigma_{qq}. \tag{A.15}$$

When B is square, it is easy to check that (A.15) coincides with the standard definition of Euclidean-norm condition number, which is $\|B\|_2\|B^{-1}\|_2$.

When B has more columns than rows, the SVD is

$$B = U\begin{bmatrix} \Sigma & 0 \end{bmatrix} V^T,$$

where U, V, and Σ have the same properties as in (A.13), while the zero block has dimension $p \times (q - p)$.

Given a real $p \times q$ matrix B (with $p \geq q$), there is a $p \times p$ orthogonal matrix Q and a $q \times q$ upper triangular matrix R such that

$$B = Q\begin{bmatrix} R \\ 0 \end{bmatrix}, \tag{A.16}$$

where the zero block has dimensions $(p - q) \times q$. The matrix R has zero elements on its diagonal if and only if B is rank deficient. Since Q is orthogonal, we can rewrite (A.16) as

$$Q^T B = \begin{bmatrix} R \\ 0 \end{bmatrix}.$$

By introducing column pivoting into the QR factorization, we can make a more explicit connection between the number of zero diagonals in R and the rank of B. There exists a $q \times q$ permutation matrix Π such that

$$B\Pi = \bar{Q} \begin{bmatrix} R_{11} & R_{12} \\ 0 & 0 \end{bmatrix}, \tag{A.17}$$

where

- $\bar{Q}$ is $p \times p$ orthogonal (not necessarily the same as Q in (A.16));
- R_{11} is $r \times r$ upper triangular with nonzero diagonal elements, where $r \leq q$ is the rank of B;
- R_{12} is $r \times (q - r)$.

In implementations of QR factorization (for example, LAPACK [9]), Π is chosen so that the diagonal elements in R_{11} appear in order of decreasing magnitude. Note that if we omit the permutation matrix Π, the matrix $\bar{Q}^T B$ still has nonzeros in the first r rows only, though these rows may not be in upper triangular form.

For further information on matrix factorizations, we refer the reader to the book by Golub and Van Loan [42], a standard reference in numerical linear algebra.

Hoffman's Lemma

A lemma of Hoffman [53] has proved to be a useful result in many areas of mathematical programming. We state it here and prove a special case.

Lemma A.3 *Let $G \in \mathbb{R}^{r \times q}$ and $H \in \mathbb{R}^{p \times q}$ be given matrices. Then there exists a nonnegative constant L depending only on G and H with the property that for all vectors*

$$\begin{bmatrix} g \\ h \end{bmatrix} \in \text{Range}\left(\begin{bmatrix} G \\ H \end{bmatrix} \right),$$

there is a solution to the system

$$Gw \leq g, \qquad Hw = h \tag{A.18}$$

that satisfies $\|w\| \leq L\|(g, h)\|$.

The result is easy to prove in the case of $r = 0$ (that is, no inequalities). By our discussions on the QR factorization, we know that there is a $p \times p$ orthogonal matrix Q such that

$$QH = \begin{bmatrix} \bar{H} \\ 0 \end{bmatrix},$$

where $\bar{H}$ has full row rank. Since $h \in \text{Range}(H)$, we also have

$$Qh = \begin{bmatrix} \bar{h} \\ 0 \end{bmatrix},$$

where $\bar{h}$ has the same number of rows as $\bar{H}$. The system (A.18) can therefore be restated equivalently as

$$\bar{H}w = \bar{h}. \tag{A.19}$$

It is trivial to verify that

$$w = \bar{H}^T(\bar{H}\bar{H}^T)^{-1}\bar{h}$$

is a solution of this system, where the matrix inverse exists because $\bar{H}$ has full row rank. Because Q is orthogonal, we have $\|\bar{h}\| = \|Qh\| = \|h\|$, and so

$$\|w\| \le \|\bar{H}^T(\bar{H}\bar{H}^T)^{-1}\|\|h\|.$$

Hence, the result of Lemma A.3 certainly holds in this special case. For the general case, we refer the reader to the original paper of Hoffman [53].

The Stewart–Todd Result and a Proof of Lemma 7.2

The following simple result, proven independently by Todd [130] and Stewart [124], has been invaluable in the asymptotic analysis of interior-point methods, where it is used to show that the affine-scaling step, calculated by setting $\sigma = 0$ in (1.12) or (1.20), has size $O(\mu)$.

The lemma refers to a class of matrices $\mathcal{D}_+$ that contains all diagonal matrices with strictly positive diagonal elements.

Lemma A.4 *Suppose that the matrix $H \in \mathbb{R}^{p \times q}$ has full row rank. Then the quantity $\chi(H)$ defined by*

$$\chi(H) = \sup_{\Sigma \in \mathcal{D}_+} \left\| \Sigma H^T (H \Sigma H^T)^{-1} \right\|$$

is finite.

Similar results have been proved by other authors; see, for example, Monteiro and Wright [98, Lemma 2.2]. The quantity $\chi(H)$ is a discontinuous function of H. For instance, the matrix

$$H(\delta) = \left[\begin{array}{cc} 1 & \delta \end{array} \right]$$

has $\chi(H(\delta)) \to \infty$ as $\delta \downarrow 0$, and yet $\chi(H(0)) = 1$. (Exercise: Calculate $\chi(H(\delta))$ explicitly.)

Lemma A.4 gives us the tool needed to prove Lemma 7.2.

Proof of Lemma 7.2. As in the proof of Lemma A.3, we do orthogonal reduction on the system $Hw = h$ to obtain an equivalent system $\bar{H}w = \bar{h}$ where $\bar{H}$ has full row rank. The least-squares problem (7.7) can then be stated equivalently as

$$\min_w \tfrac{1}{2}\|\Sigma w\|^2 \text{ subject to } \bar{H}w = \bar{h} \tag{A.20}$$

This is a convex programming problem with a strictly convex objective. Hence the vector $\bar{w}$ is a solution if and only if there is a vector $v \in \mathbb{R}^p$ such that

$$\Sigma^2 \bar{w} - \bar{H}^T v = 0, \qquad \bar{H}w = \bar{h}.$$

By substituting the first condition into the second, we obtain

$$(\bar{H}\Sigma^{-2}\bar{H}^T)v = \bar{h},$$

and hence

$$v = (\bar{H}\Sigma^{-2}\bar{H}^T)^{-1}\bar{h} \;\Rightarrow\; \bar{w} = \Sigma^{-2}\bar{H}^T(\bar{H}\Sigma^{-2}\bar{H}^T)^{-1}\bar{h}.$$

(The matrix $\bar{H}\Sigma^{-2}\bar{H}^T$ is invertible since $\bar{H}$ has full row rank and the diagonals of Σ^{-2} are strictly positive.) Hence, from Lemma A.4, we have

$$\|\bar{w}\| \le \|\Sigma^{-2}\bar{H}^T(\bar{H}\Sigma^{-2}\bar{H}^T)^{-1}\|\|\bar{h}\| \le \chi(\bar{H})\|\bar{h}\|.$$

Since $\bar{H}$ depends solely on H, it follows that $\chi(\bar{H})$ depends solely on H. The proof is completed by defining $\hat{C} = \chi(\bar{H})$. □

The Sherman–Morrison–Woodbury Formula

Often we are called on to solve a linear system in which the matrix *almost* has some nice property (such as sparsity or bandedness), except for the presence of a few awkward rows and columns or a perturbation of low rank. The Sherman–Morrison–Woodbury formula gives a way of solving the linear system by factoring only the "nice" part of the matrix and doing a few extra back-substitutions with the resulting factors to account for the awkward perturbation.

Suppose that $\tilde{G}$ is the square, nonsingular matrix that is easy to factor or invert, and let a and b be vectors. If we define G by

$$G = \tilde{G} + ab^T,$$

then, provided that G is nonsingular, we have

$$G^{-1} = \tilde{G}^{-1} - \frac{\tilde{G}^{-1}ab^T\tilde{G}^{-1}}{1 + b^T\tilde{G}^{-1}a}. \tag{A.21}$$

(To check the correctness of this formula, simply multiply this definition by $G = \tilde{G} + ab^T$, and check that this product is the identity matrix.)

The formula (A.21) can be used to find a solution of the system $Gx = d$ by solving two systems with the "easy" matrix $\tilde{G}$. We have

$$x = G^{-1}d = \tilde{G}^{-1}d - (\tilde{G}^{-1}a)\frac{b^T(\tilde{G}^{-1}d)}{1 + b^T(\tilde{G}^{-1}a)}. \tag{A.22}$$

Therefore, we need only calculate $\tilde{G}^{-1}a$ and $\tilde{G}^{-1}d$ and perform some elementary vector arithmetic to obtain the solution x.

The formula (A.21) can be extended to updates of rank greater than 1. Suppose that U and V are matrices in $\mathbb{R}^{n \times p}$ for some p between 1 and n. If we define

$$G = \tilde{G} + UV^T, \tag{A.23}$$

then

$$G^{-1} = \tilde{G}^{-1} - (I + V^T\tilde{G}^{-1}U)^{-1}\tilde{G}^{-1}UV^T\tilde{G}^{-1}. \tag{A.24}$$

To solve $Gx = d$, we need to perform $p + 1$ solves with the factors of $\tilde{G}$ (to find $\tilde{G}^{-1}d$ and $\tilde{G}^{-1}U$), and to invert the $p \times p$ matrix $(I + V^T\tilde{G}^{-1}U)$. The latter operation is inexpensive if $p \ll n$, which is usually the case in practical applications of this formula.

Asymptotic Convergence

In the asymptotic analysis of algorithms, we are often interested in the rate at which a sequence converges to its limit (see Chapters 7 and 10). We mention here some of the terminology used to describe rates of convergence.

Consider a sequence $\{\beta_k\}$ of positive scalars that converges to zero; that is,

$$\lim_{k\to\infty} \beta_k = 0.$$

The sequence is said to converge *Q-linearly* if there exists a scalar $\rho \in (0,1)$ such that

$$\frac{\beta_{k+1}}{\beta_k} \le \rho$$

for all k sufficiently large. *Q-superlinear* convergence occurs when we have

$$\lim_{k\to\infty} \frac{\beta_{k+1}}{\beta_k} = 0,$$

while the convergence is *Q-quadratic* if there is a constant C such that

$$\frac{\beta_{k+1}}{\beta_k^2} \le C$$

for all k sufficiently large. *Q-superquadratic* convergence is indicated by

$$\lim_{k\to\infty} \frac{\beta_{k+1}}{\beta_k^2} = 0.$$

It is obvious that these four terms satisfy the following relationship:

$$\text{Q-superquadratic} \Rightarrow \text{Q-quadratic} \Rightarrow \text{Q-superlinear} \Rightarrow \text{Q-linear}.$$

Generally, we say that the convergence has *Q-order p* if

$$\liminf_{k\to\infty} \frac{\log \beta_{k+1}}{\log \beta_k} \ge p.$$

It is easy to prove that for any $\delta \in (0,p)$, a sequence with the Q-order p property has

$$\lim_{k\to\infty} \frac{\beta_{k+1}}{\beta_k^{p-\delta}} = 0.$$

When $\delta = 0$, this limit may not hold; in fact, we may have $\beta_{k+1}/\beta_k^p \to \infty$. An example of this behavior is seen in Chapter 7, where we discuss a sequence that converges with Q-order 2 but not quadratically.

The prefix "Q" is sometimes understood so that terms such as "superlinear" and "quadratic" can be taken to mean Q-superlinear and Q-quadratic, respectively.

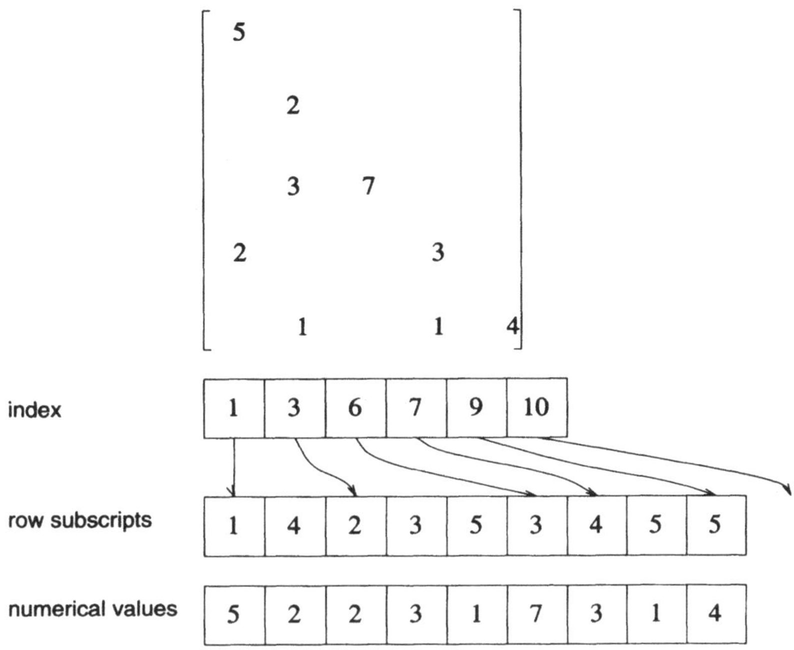

Figure A.1. Storing the lower triangle of a sparse symmetric matrix.

A slightly weaker form of convergence is denoted by the prefix "R." We say that a sequence $\{\beta_k\}$ converges *R-linearly* if it is dominated by a Q-linearly convergent sequence; that is, if there is a sequence $\{\hat{\beta}_k\}$ such that $0 \leq \beta_k \leq \hat{\beta}_k$ for all k with $\{\hat{\beta}_k\}$ converging Q-linearly. R-superlinear and R-quadratic convergence are defined analogously.

Storing Sparse Matrices

In Chapter 11 we discuss factorization of sparse symmetric matrices. Algorithms exploit sparsity by storing only the nonzero elements of the matrix, by omitting arithmetic operations that involve zero arguments, and by avoiding the creation of too many new nonzero elements during the factorization process.

To store a sparse matrix, we store all the nonzero elements alongside information about their row and column location in the matrix. A typical scheme is shown in Figure A.1. The nine elements in this sample matrix are stored in a *numerical values* vector in column-major form. This vector is indexed by two vectors of pointers. The *row subscript* vector contains the row index of the corresponding element in the numerical values vector. The *index* vector is shorter; its jth element points to the location in the other two vectors at which column j commences. (The last element of index points to the location *after* the last nonzero element.) For an $m \times m$ matrix with nnz nonzeros, the total amount of storage needed is $m + 1 + nnz$ integers and nnz real number (or double precision) locations.

When the matrix is symmetric, it is necessary only to store its lower triangle, since the upper triangle simply duplicates this information. A variant on the scheme depicted in Figure A.1 that is often used when the matrix is lower triangular or symmetric positive definite is to store the *strict* lower triangle by using the scheme above, and the diagonal elements separately, in a vector of length m.

The Turing Machine

The Turing machine, a conceptual model of computing, has just three components (see Figure A.2):

- a *tape*, made up of a number of discrete cells, each of which contains a symbol drawn from a finite alphabet. The tape is used for input, output, and storage.
- a *tape head*, which can move in both directions along the tape and read and write to its cells.
- a *finite state controller*, which adopts one of a finite number of states at each time step and controls the actions of the tape head.

Often, each cell of the tape holds a single bit of data, so the alphabet consists of just the two symbols 0 and 1. At a typical time step, the Turing machine performs the following actions:

(1) it reads the symbol under the tape head;

(2) it changes its state, the new state being determined solely by the current state and the symbol just read;

(3) it updates the cell under the tape head with a new symbol;

(4) it moves the tape head one cell to the left or right.

An algorithm controls the actions taken by the machine in steps (2), (3), and (4). Given each possible combination of current state and tape symbol from step (1), the algorithm prescribes exactly how the state is changed, how the tape cell is updated, and how the tape head is moved.

Turing machines solve a particular problem (such as linear programming) by reading an instance of the problem from the tape and, after some computation, writing the solution to the tape and terminating. In the case of linear programming, the input would be the data (A, b, c), coded in the alphabet of the Turing machine. The output could be a primal-dual solution (x^*, λ^*, s^*) or a special symbol indicating that the primal or dual (or both) is infeasible. The output must, of course, also be coded in the Turing machine alphabet.

The time required by the algorithm to solve each problem instance is simply the number of steps executed by the Turing machine before it terminates.

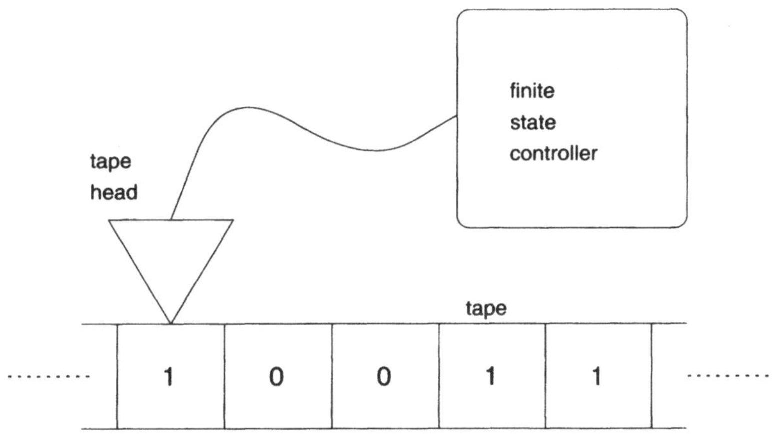

Figure A.2. The Turing machine.

The strength of the Turing machine lies in its simplicity and universality. Although far removed from the scientific/numerical computing process, it can be used as an "assembly language" on which to build more complex and useful models of computation, such as the rational number model used in Chapter 3.

Appendix B

Software Packages

The number of primal-dual codes in wide circulation has increased steadily during the past three years. We outline some of the well-known codes here. The information is current as of February 1996 but is of course subject to change. Updates will be posted on the World Wide Web at

`http://www.mcs.anl.gov/otc/InteriorPoint/`

Please contact the author to suggest additions and modifications.

Up-to-date information on interior-point research is available from *Interior-Point Methods Online*, a World Wide Web site maintained at Argonne National Laboratory at

`http://www.mcs.anl.gov/otc/Guide/faq/`

For general information about linear programming, including software information for simplex-based codes and codes for network and integer programming, see the linear programming FAQ at

`http://www.mcs.anl.gov/home/otc/Guide/faq/`

For users who wish to solve the occasional linear program and do not have the patience or the resources to download one of the codes below, the NEOS Server has a facility for remote solution through the Internet. Currently, users submit their problem as an MPS file through e-mail, anonymous ftp, the World Wide Web, or a free submission tool that runs under Xwindows. Results and runtime analysis are returned likewise. See the following URL for details:

`http://www.mcs.anl.gov/neos/Server/`

The PCx code described below is one of the software packages offered through the NEOS Server.

Most of the terms in the summaries below are explained in the text, but a few need some clarification. The term "Vertex Solution" refers to the techniques for recovering an optimal basis that we described in Chapter 7. The "Input" section in each record below indicates the ways in which users can interface their problems to the software. Some packages are callable; that is, the user can call them as a subroutine after having filled in the appropriate data structures. The main appeal of callable packages is that they are easily incorporated into larger software packages. Some packages are available via modeling language interfaces. Modeling languages such as AMPL [32] and GAMS allow users to define their models and data in intuitive terms, defining data structures, variable names, and so on in a way that naturally fits their application. The modeling language does the hard work of converting the user-defined model and data into a format acceptable to the linear programming code. After the solution is found, it inverts the process, expressing the output from this code in terms of the user's original model. More information on AMPL can be found at

`http://achille.research.att.com/ampl/`

and on GAMS at

`http://www.gams.com/`

MPS is the long-established input standard for linear programming packages. An MPS file is simply a file of ASCII text that defines the names of the primal and dual variables and specifies the objectives and constraints. Inequality and equality constraints, upper and lower bounds, free variables, are allowed in MPS files. Despite its shortcomings, the MPS format retains the advantage of universality. All solvers accept MPS input, and the standard test problems are distributed as MPS files. A thorough description of the MPS standard is given by Murtagh [99].

BPMPD

`http://www.sztaki.hu/meszaros/bpmpd/`

Language: Fortran

Algorithm: Mehrotra predictor-corrector with multiple corrections

Presolving: Yes

Input: MPS

Vertex Solution: No

Linear Algebra: Sparse Cholesky. Inexact minimum-local-fill ordering, with supernodes.

Other Information: See Mészáros [90]

Availability: Free for noncommercial use

CPLEX/Barrier `http://www.cplex.com/products/barsolv.html`

Language: C

Algorithm: Modified Mehrotra predictor-corrector, with option for Gondzio's [43] multiple central corrections

Presolving: Yes

Input: MPS, CPLEX format, AMPL, GAMS, binary file, callable

Vertex Solution: Yes, if requested

Linear Algebra: Cholesky factorization of normal equations form. Various options for ordering heuristics. Handles dense columns.

Availability: Proprietary

HOPDM
`http://www.maths.ed.ac.uk/ gondzio/software/hopdm.html`

Language: Fortran

Algorithm: Mehrotra predictor-corrector, with multiple corrections if requested

Presolving: Yes

Input: MPS, callable

Vertex Solution: No

Linear Algebra: Sparse Cholesky. Minimum-degree ordering, with supernodes and other enhancements. Handles dense columns with a Schur complement approach, as described in the Notes and References section in Chapter 11.

Other Information: See Gondzio [43, 44]

Availability: Public domain

LIPSOL `http://www.caam.rice.edu/ zhang/lipsol/`

Language: Sparse Cholesky code in Fortran, remaining code in MATLAB. User must have MATLAB installed in a UNIX environment.

Algorithm: Mehrotra predictor-corrector

Presolving: Some

Input: MPS, MATLAB binary, LPP (LP-Plain, a simple what-you-see-is-what-you-get format suitable for small problems)

Vertex Solution: No

Linear Algebra: Calls the sparse Cholesky solver of Ng and Peyton [103], (written in Fortran) and the multiple-minimum-degree ordering code of Liu [74]. Handles dense columns, using either the Sherman–Morrison–Woodbury approach described in Chapter 11 or a preconditioned conjugate gradient algorithm applied to the normal equations, with the preconditioner constructed from the sparse part of AD^2A^T (see (11.14)).

Other Information: See Zhang [163]

Availability: Public domain

LOQO `http://www.orfe.princeton.edu/ loqo`

Language: C

Algorithm: Mehrotra predictor-corrector

Presolving: No

Input: MPS, AMPL, GAMS, callable

Vertex Solution: No

Linear Algebra: Factors the augmented system matrix (11.2a) directly as

$$\begin{bmatrix} 0 & A \\ A^T & -D^{-2} \end{bmatrix} = L\hat{D}L^T,$$

where $\hat{D}$ is a diagonal matrix with both positive and negative diagonal entries (*not* a block diagonal matrix with 1×1 and 2×2 diagonal blocks, as in (11.15)). The ordering heuristic is based on minimum local fill, with a "preference" assigned to either the first m or last n columns of the augmented matrix, depending on whether A has any dense rows or dense columns. See Vanderbei [137] for details.

Special Features: Also solves convex QP problems and general convex programming problems.

Other Information: See Vanderbei [136, 138]

Availability: Free for research purposes, fee for commercial use

---

Newton Barrier XPRESS-MP `http://www.dash.co.uk/` `http://www.dashopt.com/`

Language: C and Fortran

Algorithm: Modified Mehrotra predictor-corrector (next release will be based on the homogeneous model)

Presolving: Yes

Input: MPS format, GAMS, binary file from XPRESS-MP modeller, callable subroutine library

Vertex Solution: Yes, if requested. (Method is based on Andersen and Ye [7].)

Linear Algebra: Finds a supernodal Cholesky decomposition (LDL^T) of the normal equations, using blocking for the cache memory. Minimum-degree ordering (multiple or approximate) or a variant of the inexact minimum-local-fill ordering. Handles dense columns using the modified Schur complement procedure described in [8].

Availability: Proprietary

OSL/EKKBSLV `http://www.research.ibm.com/osl/`

Language: Fortran

Algorithm: Primal barrier, Mehrotra predictor-corrector

Presolving: Offered in a separate OSL routine, EKKPRSL

Input: MPS, GAMS, callable, spreadsheet

Vertex Solution: Yes, if requested

Linear Algebra: Minimum-degree ordering heuristic with many enhancements, including supernodes. Handles dense columns by using a preconditioned conjugate gradient algorithm applied to the normal equations (11.3a), with the preconditioner constructed from the "sparse part" of AD^2A^T (see (11.14)).

Other Information: A plain primal-dual path-following algorithm (similar to Algorithm IPF from Chapter 6) is also offered as an option, but the documentation recommends the Mehrotra predictor-corrector option. See the following URLs for more specific information:

`http://www.research.ibm.com:80/osl/ekkgc6.html`
`http://www.research.ibm.com/osl/ekkgbslv.html`

Availability: Proprietary. Free benchmarking (subject to size restrictions) is available through the Internet. See the following URL for details:

`http://www.research.ibm.com/osl/bench.html`

PCx `http://www.mcs.anl.gov/otc/Tools/PCx/`

Language: Sparse Cholesky code and its interface in Fortran, remaining code in C

Algorithm: Mehrotra predictor-corrector

Presolving: Yes

Input: APML, MPL, MPS, callable. Java and Windows-95 interfaces are available; see the web site.

Vertex Solution: No

Linear Algebra: Like LIPSOL, PCx calls the sparse Cholesky solver of Ng and Peyton [103] and the multiple minimum-degree ordering codes of Liu [74], modified slightly to handle small pivots. A version for IBM RS6000 machines that calls IBM's solver WSSMP is also available. See the web page for details.

Special Features: PCx is offered as an option on the NEOS Server. Users can submit MPS files and receive results remotely over the Internet, without downloading or installing the code on their local hardware.

Other Information: See Czyzyk, Mehrotra, and Wright [23]

Availability: Freely available for research and evaluation. A licence must be obtained for production use in a commercial environment.

Bibliography

[1] I. ADLER AND R. D. C. MONTEIRO, *Limiting behavior of the affine scaling continuous trajectories for linear programming problems*, Mathematical Programming, 50 (1991), pp. 29–51.

[2] R. K. AHUJA, T. L. MAGNANTI, AND J. B. ORLIN, *Network Flows: Theory, Algorithms, and Applications*, Prentice-Hall, Englewood Cliffs, N.J., 1993.

[3] F. ALIZADEH, *Interior point methods in semidefinite programming with applications to combinatorial optimization*, SIAM Journal on Optimization, 5 (1995), pp. 13–51.

[4] F. ALIZADEH, J.-P. A. HAEBERLY, AND M. L. OVERTON, *Primal-dual interior-point methods for semidefinite programming: Convergence rates, stability, and numerical results*, Technical Report, Computer Science Department, Courant Institute of Mathematical Sciences, New York University, New York, N.Y., 1996.

[5] E. D. ANDERSEN AND K. D. ANDERSEN, *Presolving in linear programming*, Mathematical Programming, 71 (1995), pp. 221–245.

[6] E. D. ANDERSEN, J. GONDZIO, C. MÉSZÁROS, AND X. XU, *Implementation of interior-point methods for large scale linear programming*, Technical Report 1996.3, Logilab, HEC Geneva, Section of Management Studies, University of Geneva, Geneva, Switzerland, January 1996. To appear in *Interior Point Methods in Mathematical Programming*, Kluwer Academic Publishers.

[7] E. D. ANDERSEN AND Y. YE, *Combining interior-point and pivoting algorithms for linear programming*, Technical Report, Department of Management Sciences, University of Iowa, Iowa City, Ia., 1995. To appear in Management Science.

[8] K. D. ANDERSEN, *A modified Schur complement method for handling dense columns in interior-point methods for linear programming*, ACM Transactions on Mathematical Software, 22 (1996), pp. 348–356.

[9] E. ANDERSON, Z. BAI, C. BISCHOF, J. DEMMEL, J. DONGARRA, J. DU CROZ, A. GREENBAUM, S. HAMMARLING, A. MCKENNEY, S. OSTROUCHOV, AND D. SORENSEN, *LAPACK User's Guide*, SIAM, Philadelphia, 1992.

[10] M. ANITESCU, G. LESAJA, AND F. A. POTRA, *Equivalence of $P_*(\kappa)$ linear complementarity problems with respect to Newton-type algorithms*, Report on Computational Mathematics 71, Department of Mathematics, University of Iowa, Iowa City, Ia., June 1995.

[11] K. ANSTREICHER, J. JI, F. POTRA, AND Y. YE, *Average performance of a self-dual interior-point algorithm for linear programming*, in Complexity in Numerical Optimization, P. Pardalos, ed., World Scientific, River Edge, N.J., 1993, pp. 1–15.

[12] C. ASHCRAFT, R. L. GRIMES, AND J. G. LEWIS, *Accurate symmetric indefinite linear equation solvers*, SIAM Journal on Matrix Analysis and Applications, to appear.

[13] D. A. BAYER AND J. C. LAGARIAS, *The nonlinear geometry of linear programming. I. Affine and projective scaling trajectories*, Transactions of the AMS, 314 (1989), pp. 499–526.

[14] ——, *The nonlinear geometry of linear programming. II. Legendre transform coordinates and central trajectories*, Transactions of the AMS, 314 (1989), pp. 527–581.

[15] ——, *The nonlinear geometry of linear programming. III. Projective Legendre transform coordinates and Hilbert geometry*, Transactions of the AMS, 320 (1990), pp. 193–225.

[16] L. BLUM, M. SHUB, AND S. SMALE, *On a theory of computation and complexity over the real numbers: NP-completeness, recursive functions, and universal machines*, Bulletin of the AMS, 21 (1989), pp. 1–46.

[17] J. F. BONNANS AND C. C. GONZAGA, *Convergence of interior-point algorithms for the monotone linear complementarity problem*, Mathematics of Operations Research, 21 (1996), pp. 1–25.

[18] K.-H. BORGWARDT, *The Simplex Method, A Probabilistic Analysis*, Springer-Verlag, Berlin, 1987.

[19] S. BOYD, L. EL GHOUI, E. FERON, AND V. BALAKRISHNAN, *Linear Matrix Inequalities in Systems and Control Theory*, SIAM Publications, Philadelphia, 1994.

[20] T. J. CARPENTER, I. J. LUSTIG, J. M. MULVEY, AND D. F. SHANNO, *Higher-order predictor-corrector interior-point methods with application to quadratic objectives*, SIAM Journal on Optimization, 3 (1993), pp. 696–725.

[21] V. CHVÁTAL, *Linear Programming*, W. H. Freeman, New York, 1983.

[22] R. W. COTTLE, J.-S. PANG, AND R. E. STONE, *The Linear Complementarity Problem*, Academic Press, San Diego, 1992.

[23] J. CZYZYK, S. MEHROTRA, AND S. J. WRIGHT, *PCx User Guide*, Technical Report OTC 96/01, Optimization Technology Center, Argonne National Laboratory and Northwestern University, May 1996.

[24] G. B. DANTZIG, *Linear Programming and Extensions*, Princeton University Press, Princeton, N.J., 1963.

[25] D. DEN HERTOG, *Interior Point Approach to Linear, Quadratic, and Convex Programming*, Ph.D. thesis, Technische Universiteit Delft, Delft, the Netherlands, September 1992.

[26] I. I. DIKIN, *Iterative solution of problems of linear and quadratic programming*, Soviet Mathematics Doklady, 8 (1967), pp. 674–675.

[27] I. S. DUFF, *The solution of augmented systems*, in Numerical Analysis 1993, D. F. Griffiths and G. A. Watson, eds., Longman Scientific and Technical, Essex, U.K., pp. 40–55.

[28] A. S. EL-BAKRY, R. A. TAPIA, AND Y. ZHANG, *A study of indicators for identifying zero variables in interior-point methods*, SIAM Review, 36 (1994), pp. 45–72.

[29] A. V. FIACCO AND G. P. MCCORMICK, *Nonlinear Programming: Sequential Unconstrained Minimization Techniques*, Wiley, New York, 1968. Reprinted by SIAM Publications, 1990.

[30] R. FLETCHER, *Practical Methods of Optimization*, John Wiley and Sons, New York, second ed., 1987.

[31] A. FORSGREN, P. GILL, AND J. SHINNERL, *Stability of symmetric ill-conditioned systems arising in interior methods for constrained optimization*, SIAM Journal on Matrix Analysis and Applications, 17 (1996), pp. 187–211.

[32] R. FOURER, D. M. GAY, AND B. W. KERNIGHAN, *AMPL: A Modeling Language for Mathematical Programming*, The Scientific Press, South San Francisco, Calif., 1993.

[33] R. FOURER AND S. MEHROTRA, *Solving symmetric indefinite systems in an interior-point method for linear programming*, Mathematical Programming, 62 (1993), pp. 15–39.

[34] R. FREUND, *Polynomial-time algorithms for linear programming based only on primal scaling and projected gradients of a potential function*, Mathematical Programming, 51 (1991), pp. 203–222.

[35] R. W. FREUND AND F. JARRE, *A QMR-based interior-point algorithm for solving linear programs*, Numerical Analysis Manuscript 94–19, AT&T Bell Laboratories, Murray Hill, N.J., December 1994.

[36] K. R. FRISCH, *The logarithmic potential method of convex programming*, Technical Report, University Institute of Economics, Oslo, Norway, 1955.

[37] M. R. GAREY AND D. S. JOHNSON, *Computers and Intractability*, W. H. Freeman, New York, 1979.

[38] A. GEORGE AND J. W.-H. LIU, *Computer Solution of Large Sparse Positive Definite Systems*, Prentice-Hall, Englewood Cliffs, N.J., 1981.

[39] A. GEORGE AND J. W.-H. LIU, *The evolution of the minimum degree ordering*, SIAM Review, 31 (1989), pp. 1–19.

[40] P. E. GILL, W. MURRAY, M. A. SAUNDERS, J. A. TOMLIN, AND M. H. WRIGHT, *On projected Newton barrier methods for linear programming and an equivalence to Karmarkar's projective method*, Mathematical Programming, 36 (1986), pp. 183–209.

[41] A. J. GOLDMAN AND A. W. TUCKER, *Theory of linear programming*, in Linear Equalities and Related Systems, H. W. Kuhn and A. W. Tucker, eds., Princeton University Press, Princeton, N.J., 1956, pp. 53–97.

[42] G. H. GOLUB AND C. F. VAN LOAN, *Matrix Computations*, The Johns Hopkins University Press, Baltimore, Md., 2nd ed., 1989.

[43] J. GONDZIO, *Multiple centrality corrections in a primal-dual method for linear programming*, Computational Optimization and Applications, 6 (1996), pp. 137–156.

[44] ——, *HOPDM* (*version* 2.12): *A fast LP solver based on a primal-dual interior point method*, European Journal of Operations Research, 85 (1995), pp. 221–225.

[45] C. GONZAGA, *Polynomial affine algorithms for linear programming*, Mathematical Programming, 49 (1990), pp. 7–21.

[46] ——, *Path-following methods in linear programming*, SIAM Review, 34 (1992), pp. 167–224.

[47] C. C. GONZAGA AND M. J. TODD, *An $O(\sqrt{n}L)$-iteration large-step primal-dual affine algorithm for linear programming*, SIAM Journal on Optimization, 2 (1992), pp. 349–359.

[48] O. GÜLER, *Hyperbolic polynomials and interior-point methods for convex programming*, Technical Report TR95–40, Mathematics Department, University of Maryland, Baltimore County, Baltimore, Md., 1995.

[49] O. GÜLER, C. ROOS, T. TERLAKY, AND J.-P. VIAL, *A survey of the implications of the behavior of the central path for the duality theory of linear programming*, Management Science, 41 (1995), pp. 1922–1934.

[50] O. GÜLER AND Y. YE, *Convergence behavior of interior-point algorithms*, Mathematical Programming, 60 (1993), pp. 215–228.

[51] C. HELMBERG, F. RENDL, R. J. VANDERBEI, AND H. WOLKOWICZ, *An interior-point method for semidefinite programming*, SIAM Journal on Optimization, 6 (1996), pp. 342–361.

[52] N. J. HIGHAM, *Accuracy and Stability of Numerical Algorithms*, SIAM Publications, Philadelphia, 1996.

[53] A. J. HOFFMAN, *On approximate solutions of systems of linear inequalities*, Journal of Research of the National Bureau of Standards, 49 (1952), pp. 263–265.

[54] B. JANSEN, C. ROOS, AND T. TERLAKY, *A polynomial primal-dual Dikin-type algorithm for linear programming*, Mathematics of Operations Research, 21 (1996), pp. 341–353.

[55] F. JARRE AND M. SAUNDERS, *A practical interior-point method for convex programming*, SIAM Journal on Optimization, 5 (1995), pp. 149–171.

[56] L. V. KANTOROVICH, *Functional analysis and applied mathematics*, Uspekhi Matematicheskikh Nauk, 3 (1948), pp. 89–185. Translated by C. Benster as National Bureau of Standards Report 1509, Washington, D.C., 1952.

[57] N. KARMARKAR, *A new polynomial-time algorithm for linear programming*, Combinatorica, 4 (1984), pp. 373–395.

[58] N. KARMARKAR, J. C. LAGARIAS, L. SLUTSMAN, AND P. WANG, *Power series variants of Karmarkar-type algorithms*, AT&T Technical Journal, 68 (1989), pp. 20–36.

[59] G. KARYPIS, A. GUPTA, AND V. KUMAR, *A parallel formulation of interior-point algorithms*, Technical Report 94–20, Computer Science Department, University of Minnesota, Minneapolis, Minn., 1994.

[60] L. G. KHACHIYAN, *A polynomial algorithm in linear programming*, Soviet Mathematics Doklady, 20 (1979), pp. 191–194.

[61] V. KLEE AND G. J. MINTY, *How good is the simplex algorithm?*, in Inequalities, O. Shisha, ed., Academic Press, New York, 1972, pp. 159–175.

[62] M. KOJIMA, *Basic lemmas in polynomial-time infeasible-interior-point methods for linear programs*, Annals of Operations Research, 62 (1996), pp. 1–28.

[63] M. KOJIMA, N. MEGIDDO, AND S. MIZUNO, *A general framework of continuation methods for complementarity problems*, Mathematics of Operations Research, 18 (1993), pp. 945–963.

[64] ——, *A primal-dual infeasible-interior-point point algorithm for linear programming*, Mathematical Programming, Series A, 61 (1993), pp. 261–280.

[65] M. KOJIMA, N. MEGIDDO, T. NOMA, AND A. YOSHISE, *A unified approach to interior-point algorithms for linear complementarity problems*, Lecture Notes in Computer Science 538, Springer-Verlag, Berlin, 1991.

[66] M. KOJIMA, S. MIZUNO, AND A. YOSHISE, *A polynomial-time algorithm for a class of linear complementarity problems*, Mathematical Programming, 44 (1989), pp. 1–26.

[67] ——, *A primal-dual interior point algorithm for linear programming*, in Progress in Mathematical Programming: Interior-Point and Related Methods, N. Megiddo, ed., Springer-Verlag, New York, 1989, ch. 2, pp. 29–47.

[68] ——, *An $O(\sqrt{n}L)$ iteration potential reduction algorithm for linear complementarity problems*, Mathematical Programming, 50 (1991), pp. 331–342.

[69] M. KOJIMA, T. NOMA, AND A. YOSHISE, *Global convergence in infeasible-interior-point algorithms*, Mathematical Programming, 65 (1994), pp. 43–72.

[70] M. KOJIMA, M. SHIDA, AND S. SHINDOH, *Global and local convergence of predictor-corrector infeasible-interior-point algorithms for semidefinite programs*, Research Report B–305, Department of Mathematical and Computing Sciences, Tokyo Institute of Technology, Tokyo, Japan, October 1995.

[71] M. KOJIMA, S. SHINDOH, AND S. HARA, *Interior-point methods for the monotone linear complementarity problem in symmetric matrices*, SIAM Journal on Optimization, 7 (1997), to appear.

[72] C. L. LAWSON AND R. J. HANSON, *Solving Least Squares Problems*, SIAM, Philadelphia, 1995.

[73] A. S. LEWIS AND M. L. OVERTON, *Eigenvalue optimization*, in Acta Numerica 1996, Cambridge University Press, Cambridge, U.K., 1996, pp. 149–190.

[74] J. W.-H. LIU, *Modification of the minimum degree algorithm by multiple elimination*, ACM Transactions on Mathematical Software, 11 (1985), pp. 141–153.

[75] D. LUENBERGER, *Introduction to Linear and Nonlinear Programming*, Addison-Wesley, Reading, Mass., 2nd ed., 1984.

[76] I. J. LUSTIG, *Feasibility issues in primal-dual interior-point method for linear programming*, Mathematical Programming, 49 (1991), pp. 145–162.

[77] I. J. LUSTIG, R. E. MARSTEN, AND D. F. SHANNO, *Computational experience with a primal-dual interior point method for linear programming*, Linear Algebra and Its Applications, 152 (1991), pp. 191–222.

[78] ——, *Computational experience with a globally convergent primal-dual predictor-corrector algorithm for linear programming*, Mathematical Programming, 66 (1994), pp. 123–135.

[79] ——, *Interior-point methods for linear programming: Computational state of the art*, ORSA Journal on Computing, 6 (1994), pp. 1–14.

[80] I. J. LUSTIG AND E. ROTHBERG, *Gigaflops in linear programming*, Operations Research Letters, 18 (1996), pp. 157–165.

[81] O. L. MANGASARIAN, *Nonlinear Programming*, McGraw-Hill, New York, 1969.

[82] ——, *Error bounds for nondegenerate monotone linear complementarity problems*, Mathematical Programming, 48 (1990), pp. 437–445.

[83] L. MCLINDEN, *An analogue of Moreau's proximation theorem, with applications to the nonlinear complementarity problem*, Pacific Journal of Mathematics, 88 (1980), pp. 101–161.

[84] K. A. MCSHANE, C. L. MONMA, AND D. F. SHANNO, *An implementation of a primal-dual interior-point method for linear programming*, ORSA Journal on Computing, 1 (1989), pp. 70–83.

[85] N. MEGIDDO, *Pathways to the optimal set in linear programming*, in Progress in Mathematical Programming: Interior-Point and Related Methods, N. Megiddo, ed., Springer-Verlag, New York, 1989, ch. 8, pp. 131–158.

[86] ——, *On finding primal- and dual-optimal bases*, ORSA Journal on Computing, 3 (1991), pp. 63–65.

[87] S. MEHROTRA, *Higher order methods and their performance*, Technical Report 90–16R1, Department of Industrial Engineering and Management Sciences, Northwestern University, Evanston, Ill., 1991.

[88] ——, *On the implementation of a primal-dual interior point method*, SIAM Journal on Optimization, 2 (1992), pp. 575–601.

[89] S. MEHROTRA AND J.-S. WANG, *Conjugate gradient based implementation of interior point methods for network flow problems*, Technical Report 95–70, Department of Industrial Engineering and Management Sciences, Northwestern University, Evanston, Ill., October 1995.

[90] C. MÉSZÁROS, *The inexact minimum local fill-in ordering algorithm*, Working Paper 95–7, Computer and Automation Institute, Hungarian Academy of Sciences, Budapest, Hungary, 1995.

[91] S. MIZUNO, M. KOJIMA, AND M. TODD, *Infeasible-interior-point primal-dual potential-reduction algorithms for linear programming*, SIAM Journal on Optimization, 5 (1995), pp. 52–67.

[92] S. MIZUNO, M. TODD, AND Y. YE, *On adaptive step primal-dual interior-point algorithms for linear programming*, Mathematics of Operations Research, 18 (1993), pp. 964–981.

[93] ——, *A surface of analytic centers and primal–dual infeasible-interior–point algorithms for linear programming*, Mathematics of Operations Research, 20 (1995), pp. 135–162.

[94] R. D. C. MONTEIRO AND I. ADLER, *Interior path-following primal-dual algorithms. Part* I: *Linear programming*, Mathematical Programming, 44 (1989), pp. 27–41.

[95] R. D. C. MONTEIRO, I. ADLER, AND M. G. C. RESENDE, *A polynomial–time primal–dual affine scaling algorithm for linear and convex quadratic programming and its power series extension*, Mathematics of Operations Research, 15 (1990), pp. 191–214.

convex quadratic programming and its power series extension, Mathematics of Operations Research, 15 (1990), pp. 191–214.

[96] R. D. C. Monteiro, T. Tsuchiya, and Y. Wang, *A simplified global convergence proof of the affine scaling algorithm*, Annals of Operations Research, 47 (1993), pp. 443–482.

[97] R. D. C. Monteiro and S. J. Wright, *Local convergence of interior-point algorithms for degenerate monotone LCP*, Computational Optimization and Applications, 3 (1994), pp. 131–155.

[98] ———, *Superlinear primal-dual affine scaling algorithms for LCP*, Mathematical Programming, 69 (1995), pp. 311–333.

[99] B. A. Murtagh, *Advanced Linear Programming: Computation and Practice*, McGraw-Hill, New York, 1981.

[100] G. L. Nemhauser and L. A. Wolsey, *Integer and Combinatorial Optimization*, John Wiley and Sons, New York, 1988.

[101] Y. E. Nesterov and A. S. Nemirovskii, *Interior Point Polynomial Methods in Convex Programming*, SIAM Publications, Philadelphia, 1994.

[102] Y. E. Nesterov and M. Todd, *Primal-dual interior-point methods for self-scaled cones*, Technical Report 1125, School of Operations Research and Industrial Engineering, Cornell University, Ithaca, N.Y., August 1995.

[103] E. Ng and B. W. Peyton, *Block sparse Cholesky algorithms on advanced uniprocessor computers*, SIAM Journal on Scientific Computing, 14 (1993), pp. 1034–1056.

[104] J. Nocedal and S. J. Wright, *Numerical Optimization*, forthcoming, 1997.

[105] C. C. Paige and M. A. Saunders, *LSQR: an algorithm for sparse linear equations and sparse least squares*, ACM Transactions on Mathematical Software, 8 (1982), pp. 43–71.

[106] D. B. Ponceleón, *Barrier methods for large-scale quadratic programming*, Ph.D. thesis, Stanford University, 1990.

[107] L. F. PORTUGAL, M. G. C. RESENDE, J. VEIGA, AND J. J. JÚDICE, *A truncated primal-feasible dual-infeasible interior-point network flow method*, Technical Report, Mathematical Sciences Research Center, AT&T Bell Laboratories, Murray Hill, N.J., 1994.

[108] F. A. POTRA, *A quadratically convergent predictor-corrector method for solving linear programs from infeasible starting points*, Mathematical Programming, 67 (1994), pp. 383–406.

[109] F. A. POTRA AND R. SHENG, *A large-step infeasible-interior-point method for the P_* matrix LCP*, SIAM Journal on Optimization, 7 (1997), to appear.

[110] ——, *A superlinearly convergent primal-dual infeasible-interior-point algorithm for semidefinite programming*, Report on Computational Mathematics 78/1995, Department of Mathematics, University of Iowa, Iowa City, Ia., October 1995.

[111] D. RALPH AND S. J. WRIGHT, *Superlinear convergence of an interior-point method for monotone variational inequalities*, in Complementarity and Variational Problems: State of the Art, M. C. Ferris and J.-S. Pang, SIAM Publications, Philadelphia, forthcoming.

[112] J. RENEGAR, *A polynomial-time algorithm, based on Newton's method, for linear programming*, Mathematical Programming, 40 (1988), pp. 59–93.

[113] ——, *Condition numbers, the barrier method, and the conjugate gradient method*, SIAM Journal on Optimization, 6 (1996), pp. 879–912.

[114] ——, *Some perturbation theory for linear programming*, Mathematical Programming, 65 (1994), pp. 73–92.

[115] ——, *Incorporating condition measures into the complexity theory of linear programming*, SIAM Journal on Optimization, 5 (1995), pp. 506–524.

[116] R. T. ROCKAFELLAR, *Convex Analysis*, Princeton University Press, Princeton, N.J., 1970.

[117] D. J. ROSE AND R. E. TARJAN, *Algorithmic aspects of vertex elimination on graphs*, in Proceedings of the 7th Annual Symposium on the

Theory of Computing, Association for Computing Machinery, New York, 1975, pp. 245–254.

[118] E. ROTHBERG AND B. HENDRICKSON, *Sparse matrix ordering methods for interior-point linear programming*, Technical Report SAND96-0475J, Sandia National Laboratory, 1996.

[119] A. SCHRIJVER, *Theory of Linear and Integer Programming*, John Wiley and Sons, New York, 1986.

[120] V. E. SHAMANSKII, *Ob odnoj modifikatsii metody Newtona (on a modification of Newton's method)*, Ukrainskii Matematichnii Zhurnal, 19 (1967), pp. 133–138.

[121] G. SONNEVEND, *An "analytic center" for polyhedrons and new classes of global algorithms for linear (smooth, convex) programming*, in System Modelling and Optimization: Proceedings of the 12th IFIP Conference held in Budapest, Hungary, September, 1985, A. Prekopa, J. Szelezsan, and B. Strazicky, eds., Lecture Notes in Control and Information Sciences 84, Springer-Verlag, Berlin, 1986, pp. 866–876.

[122] G. SONNEVEND, J. STOER, AND G. ZHAO, *On the complexity of following the central path of linear programs by linear extrapolation*, Methods of Operations Research, 62 (1989), pp. 19–31.

[123] ——, *On the complexity of following the central path of linear programs by linear extrapolation* II, Mathematical Programming, 52 (1991), pp. 527–553.

[124] G. W. STEWART, *On scaled projections and pseudoinverses*, Linear Algebra and Its Applications, 112 (1989), pp. 189–193.

[125] K. TANABE, *Centered Newton method for mathematical programming*, in System Modeling and Optimization: Proceedings of the 13th IFIP conference, Lecture Notes in Control and Information Systems 113, Berlin, August/September 1987, Springer-Verlag, New York, 1988, pp. 197–206.

[126] R. A. TAPIA, Y. ZHANG, AND Y. YE, *On the convergence of the iteration sequence in primal-dual interior-point methods*, Mathematical Programming, 68 (1995), pp. 141–154.

[127] E. TARDOS, *A strongly polynomial algorithm to solve combinatorial linear programs*, Operations Research, 34 (1986), pp. 250–256.

[128] M. TODD, *Potential reduction methods in mathematical programming*, Technical Report 1112, Cornell University, Ithaca, N.Y., 1995. To appear in Mathematical Programming, Series B.

[129] M. TODD AND B. P. BURRELL, *An extension of Karmarkar's algorithm for linear programming using dual variables*, Algorithmica, 1 (1986), pp. 409–424.

[130] M. J. TODD, *A Dantzig-Wolfe-like variant of Karmarkar's interior-point linear programming algorithm*, Operations Research, 38 (1990), pp. 1006–1018.

[131] M. J. TODD AND Y. YE, *A centered projective algorithm for linear programming*, Mathematics of Operations Research, 15 (1990), pp. 508–529.

[132] J. F. TRAUB, *Iterative Methods for the Solution of Equations*, Prentice-Hall, Englewood Cliffs, N.J., 1964.

[133] P. TSENG AND Z.-Q. LUO, *On the convergence of the affine–scaling algorithm*, Mathematical Programming, 56 (1992), pp. 301–319.

[134] T. TSUCHIYA AND M. MURAMATSU, *Global convergence of a long-step affine scaling algorithm for degenerate linear programming problems*, SIAM Journal on Optimization, 5 (1995), pp. 525–551.

[135] L. VANDENBERGHE AND S. BOYD, *Semidefinite programming*, SIAM Review, 38 (1996), pp. 49–95.

[136] R. J. VANDERBEI, *LOQO: An interior-point code for quadratic programming*, Technical Report SOR-94-15, Department of Civil Engineering and Operations Research, Princeton University, Princeton, N.J., 1994.

[137] ——, *Symmetric quasidefinite matrices*, SIAM Journal on Optimization, 5 (1995), pp. 100–113.

[138] ——, *LOQO User's Manual, Version* 2.21, Technical Report SOR-96-07, Department of Civil Engineering and Operations Research, Princeton University, Princeton, N.J., 1996.

[139] S. A. Vavasis, *Nonlinear Optimization*, Oxford University Press, New York, Oxford, 1991.

[140] ——, *Stable numerical algorithms for equilibrium systems*, SIAM Journal on Matrix Analysis and Applications, 15 (1994), pp. 1108–1131.

[141] S. A. Vavasis and Y. Ye, *A primal dual interior point method whose running time depends only on the constraint matrix*, Mathematical Programming, Series A, 74 (1996), pp. 79–120.

[142] J. Vera, *On the complexity of linear programming under finite precision arithmetic*, Technical Report, Department of Industrial Engineering, University of Chile, Santiago, Chile, 1994. Revised, January 1996.

[143] J. H. Wilkinson, *The Algebraic Eigenvalue Problem*, Oxford University Press, London, 1965.

[144] ——, *A priori error analysis of algebraic processes*, in Proceedings of the International Congress of Mathematics, Moscow, 1968, Izdatelstvo Mir, pp. 629–639.

[145] M. H. Wright, *Interior methods for constrained optimization*, in Acta Numerica 1992, Cambridge University Press, Cambridge, U.K., 1992, pp. 341–407.

[146] S. J. Wright, *A path-following interior-point algorithm for linear and quadratic problems*, Annals of Operations Research, 62 (1996), pp. 103–130.

[147] ——, *An infeasible-interior-point algorithm for linear complementarity problems*, Mathematical Programming, 67 (1994), pp. 29–52.

[148] ——, *Stability of augmented system factorizations in interior-point methods*, SIAM Journal on Matrix Analysis and Applications, 18 (1997), to appear.

[149] ——, *Stability of linear equations solvers in interior-point methods*, SIAM Journal on Matrix Analysis and Applications, 16 (1994), pp. 1287–1307.

[150] ——, *Modified Cholesky factorizations in interior-point algorithms for linear programming*, Preprint ANL/MCS-P600-0596, Mathematics and

Computer Science Division, Argonne National Laboratory, Argonne, Ill., May 1996.

[151] S. J. WRIGHT AND Y. ZHANG, *A superquadratic infeasible-interior-point method for linear complementarity problems*, Mathematical Programming, 73 (1996), pp. 269–289.

[152] X. XU, *An $O(\sqrt{n}L)$-iteration large-step infeasible path-following algorithm for linear programming*, Technical Report, Department of Management Sciences, University of Iowa, Iowa City, Ia., 1994.

[153] X. XU, P. HUNG, AND Y. YE, *A simplified homogeneous and self-dual linear programming algorithm and its implementation*, Annals of Operations Research, 62 (1996), pp. 151–172.

[154] Y. YE, *An $O(n^3L)$ potential reduction algorithm for linear programming*, Mathematical Programming, 50 (1991), pp. 239–258.

[155] ——, *On the finite convergence of interior-point algorithms for linear programming*, Mathematical Programming, 57 (1992), pp. 325–336.

[156] ——, *Interior-Point Algorithm: Theory and Analysis*, John Wiley and Sons, New York, 1997, to appear.

[157] Y. YE AND K. ANSTREICHER, *On quadratic and $O(\sqrt{n}L)$ convergence of a predictor-corrector algorithm for LCP*, Mathematical Programming, Series A, 62 (1993), pp. 537–551.

[158] Y. YE, O. GÜLER, R. A. TAPIA, AND Y. ZHANG, *A quadratically convergent $O(\sqrt{n}L)$-iteration algorithm for linear programming*, Mathematical Programming, Series A, 59 (1993), pp. 151–162.

[159] Y. YE, M. J. TODD, AND S. MIZUNO, *An $O(\sqrt{n}L)$-iteration homogeneous and self-dual linear programming algorithm*, Mathematics of Operations Research, 19 (1994), pp. 53–67.

[160] Y. ZHANG, *On the convergence of a class of infeasible-interior-point methods for the horizontal linear complementarity problem*, SIAM Journal on Optimization, 4 (1994), pp. 208–227.

[161] ——, *On extending primal-dual interior-point algorithms from linear programming to semidefinite programming*, Technical Report TR95–20, Department of Mathematics and Statistics, University of Maryland, Baltimore County, Baltimore, Md., November 1995.

[162] ——, *User's guide to LIPSOL*, Department of Mathematics and Statistics, University of Maryland, Baltimore County, Baltimore, Md., July 1995.

[163] ——, *Solving large-scale linear programs by interior-point methods under the MATLAB enviroment*, Technical Report TR96-01, Department of Mathematics and Statistics, University of Maryland, Baltimore County, Baltimore, Md., 1996.

[164] Y. ZHANG, R. A. TAPIA, AND J. E. DENNIS, *On the superlinear and quadratic convergence of primal-dual interior point linear programming algorithms*, SIAM Journal on Optimization, 2 (1992), pp. 304–324.

Index